Grades
4-5

BUILDING

MATH SKILLS

I0499695

BY Brian Rhee

Over 2100 Practice Problems

Detailed Solutions

Legal Notice

Copyright © 2018 by Solomon Academy
Published by: Solomon Academy
First Edition
ISBN-13: 978-1717065520
ISBN-10: 171706552X

About This Book

This book is designed to help students build basic arithmetic and math skills. There are resources pages in the beginning of the book that illustrate how to do each type of problems. It is strongly recommended to go over the resources pages before solving practice problems.

This book contains 12 lessons with detailed solutions. Each lesson has ten practice worksheets which provides challenges to improve and strengthen students' math skills. Completing 12 lessons enables students to build their confidence and master their math skills.

About Author

Brian(Yeon) Rhee obtained a Masters of Arts Degree in Statistics at Columbia University, NY. He served as the Mathematical Statistician at the Bureau of Labor Statistics, DC. He is the Head Academic Director at Solomon Academy due to his devotion to the community coupled with his passion for teaching. His mission is to help students of all confidence level excel in academia to build a strong foundation in character, knowledge, and wisdom. Now, Solomon academy is known as the best academy specialized in Math in Northern Virginia.

Brian Rhee has published more than ten books. The titles of his books are 7 full-length practice tests for the AP Calculus AB/BC Multiple choice sections, AP Calculus, SAT 1 Math, SAT 2 Math level 2, 12 full-length practice tests for the SAT 2 Math Level 2, SHSAT/TJHSST Math workbook, and IAAT (Iowa Algebra Aptitude Test) Volume 1 and 2, CogAT form 7 Level 8, and NNAT 2 Level B Grade 1. He's currently working on other math books which will be introduced in the near future.

Brian Rhee has more than twenty years of teaching experience in math. He has been one of the most popular tutors among TJHSST (Thomas Jefferson High School For Science and Technology) students. Currently, he is developing many online math courses with www.masterprep.net for AP Calculus AB and BC, SAT 2 Math level 2 test, and other various math subjects.

SOLOMON ACADEMY

Solomon Academy is a prestigious institution of learning with numerous qualified teachers of various fields of education. Our mission is to thoroughly teach students of all ages and confidence levels, elevate skills to the highest standard of education, and provide them with all the tools and materials to succeed.

5723 Centre Square Drive
Centreville, VA 20120
Tel: 703-988-0019

Email: solomonacademyva@gmail.com
info@solomonacademy.net

CLASSES OFFERED

MATHEMATICS	TESTING	ENGLISH
1st-6th grade math	CogAt	1st-6th Reading
Algebra 1, 2	IAAT and SOL 7	1st-6th Writing
Geometry	TJHSST Prep	Essay Writing
Pre-Calculus	SAT/ACT Prep	SAT Writing
AP Calculus AB BC	SAT 2 Subject Tests	
AP Statistics	MathCounts	
Multivariate Calculus	AMC 10/12	

LEARN FROM THE AUTHOR

Private sessions with Brian Rhee is also available on the following subjects: SAT Math, SAT 2 Subject Math Level 2, Pre-Calculus, AP Calculus AB/BC, AP Statistics, IB SL/HL, Multivariate Calculus, Linear Algebra, AMC 8/10/12, and AIME.

Feel free to contact me at solomonacademyva@gmail.com

Acknowledgements

I wish to acknowledge my deepest appreciation to my wife, Sookyung, who has continuously given me wholehearted support, encouragement, and love. Without you, I could not have completed this book.

Thank you to my sons, Joshua and Jason, who have given me big smiles and inspiration. I love you all.

Thank you to Mr. Kwon from www.Masterprep.net, who has given me opportunities to develop online math courses for various math subjects.

Contents

How to do multiplication and division?

- Multiplication without regrouping

    1. Multiply 1×3.
    2. Multiply 2×3.
    3. Multiply 3×3.

$$\begin{array}{r} 3\ 2\ 1 \\ \times \qquad 3 \\ \hline \end{array} \longrightarrow \begin{array}{r} 3\ 2\ 1 \\ \times \qquad 3 \\ \hline 9\ 6\ 3 \end{array}$$

$9\ 6\ 3 {-}1 \times 3$

$\qquad 2 \times 3$

$\qquad 3 \times 3$

- Multiplication with regrouping

    1. Multiply 47×6.
    2. Multiply 47×5.
    3. Add.

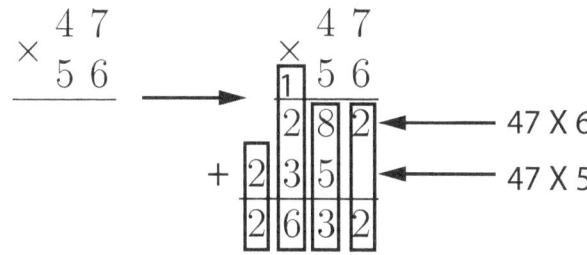

$$\begin{array}{r} 4\ 7 \\ \times \qquad 5\ 6 \\ \hline \end{array} \longrightarrow \begin{array}{r} 4\ 7 \\ \times \qquad 5\ 6 \\ \hline \end{array}$$

47×6

47×5

- Division without a remainder

    1. 3 goes into 4 one time.

    2. Subtract 3 from 4.

    3. Bring down the 2. 3 goes into 12 four times.

$$
3\overline{)42} \longrightarrow
\begin{array}{r}
14 \\
3\overline{)42} \\
30 \\
\hline
12 \\
12 \\
\hline
0
\end{array}
$$

- Division with a remainder

    1. 5 goes into 6 one time.

    2. Subtract 5 from 6.

    3. Bring down the 7. 5 goes into 17 three times.

    4. The remainder is 2.

$$
5\overline{)67} \longrightarrow
\begin{array}{r}
13 \\
5\overline{)67} \\
50 \\
\hline
17 \\
15 \\
\hline
2
\end{array}
$$

Fractions

- Changing fractions to simplest form

    1. Divide the numerator and the denominator by a common factor, a number that can divide the numerator and the denominator exactly.

    2. Keep dividing until the numerator and the denominator cannot be smaller.

$$\frac{12}{18} = \frac{12 \div 2}{18 \div 2} = \frac{6 \div 3}{9 \div 3} = \frac{2}{3}$$

- Adding or subtracting fractions with the same denominators

    1. Add or subtract the numerators and keep the denominators the same.

    2. Write the result to simplest form.

$$\frac{2}{6} + \frac{1}{6} = \frac{2+1}{6} = \frac{3}{6} = \frac{1}{2}, \qquad \frac{5}{8} - \frac{3}{8} = \frac{5-3}{8} = \frac{2}{8} = \frac{1}{4}$$

- Multiplying fractions

  1. Multiply the numerators and then multiply the denominators.

  2. Write the result to simplest form.

$$\frac{2}{3} \times \frac{3}{4} = \frac{2 \times 3}{3 \times 4} = \frac{6}{12} = \frac{1}{2}$$

- Dividing fractions

  1. Turn the second fraction upside down(the reciprocal of the second fraction).

  2. Multiply the first fraction by the reciprocal of the second fraction

  3. Write the result to simplest form.

$$\frac{3}{4} \div \frac{5}{2} = \frac{3}{4} \times \frac{2}{5} = \frac{3 \times 2}{4 \times 5} = \frac{6}{20} = \frac{3}{10}$$

Decimals

- Changing fractions to decimals

  1. Find a number you can multiply by the denominator to make it 10, or 100, or 1000.
  2. Multiply both numerator and denominator.
  3. Write down the numerator and put the decimal point in the correct spot.

$$\frac{3}{4} = \frac{3 \times 25}{4 \times 25} = \frac{75}{100} = 0.75$$

- Changing decimals to fractions

  1. Put the numbers to the right of the decimal point in the numerator.
  2. The denominator is either 10, or 100, or 1000 depending on the number of digits after the decimal point.
  3. Write the result to simplest form.

$$0.2 = \frac{2}{10} = \frac{1}{5}, \qquad 0.25 = \frac{25}{100} = \frac{1}{4}, \qquad 0.125 = \frac{125}{1000} = \frac{1}{8}$$

- Adding decimals

    1. When putting the numbers in a vertical column, align the decimal points.

    2. Add each column of digits starting on the right and working left.

    3. Place the decimal point in the answer directly below the decimal points.

$$
\begin{array}{r}
{\scriptstyle 1\ \ 1} \\
1.3\,4 \\
+\ 0.7\,8 \\
\hline
2.1\,2
\end{array}
$$

- Subtracting decimals

    1. When putting the numbers in a vertical column, align the decimal points.

    2. Subtract each column of digits starting on the right and working left.

    3. Place the decimal point in the answer directly below the decimal points.

$$
\begin{array}{r}
2.1\,2 \\
-\ 1.2\,8 \\
\hline
0.8\,4
\end{array}
$$

- Multiplying decimals

  1. Put two decimals in a vertical column.
  2. Staring on the right, multiply each digit in the top number by each digit in the bottom number, just as with whole numbers.
  3. Add the products.
  4. Place the decimal point in the answer by staring at the right and moving a number of places equal to the sum of the decimal places in both numbers multiplied.

$$
\begin{array}{r}
2.4 \\
\times \quad 3.6 \\
\hline
1\,4\,4 \\
7\,2 \quad \\
\hline
8.6\,4
\end{array}
$$

     2.4   (1 decimal place)
× 3.6   (1 decimal place)

8.6 4   (2 decimal places)

- Dividing decimals

  1. If the divisor(1.2) is not a whole number, move decimal point to right to make it a whole number.
  2. Move the decimal point in dividend(3.24) the same number of places.
  3. Divide as usual.
  4. Put decimal point directly above decimal point in the dividend.

$$
1.2\,\overline{)\,3.24} \quad \longrightarrow \quad 12\,\overline{)\,32.4}
$$

$$
\begin{array}{r}
2.7 \\
12\,\overline{)\,32.4} \\
24 \\
\hline
84 \\
84 \\
\hline
0
\end{array}
$$

1. × 1 0 3 3
        2 5

2. × 1 3 4 4
        1 4

3. × 1 6 2 9
        5 3

4. × 1 8 5 6
        1 6

5. × 1 5 2 8
        4 6

6. × 1 4 3 2
        1 8

7. × 1 6 6 1
        3 5

8. × 1 9 6 5
        3 2

9. × 1 2 8 8
        4 9

10. × 1 1 4 5
        1 9

11. × 1 6 8 5
        1 2

12. × 1 8 5 7
        1 3

1. $\times$ 2 8 4 3
          2 8

2. $\times$ 1 5 3 2
          4 5

3. $\times$ 2 4 5 1
          2 1

4. $\times$ 1 2 7 8
          3 7

5. $\times$ 2 3 8 0
          1 9

6. $\times$ 2 8 2 6
          3 3

7. $\times$ 2 1 2 3
          2 5

8. $\times$ 1 5 8 3
          5 5

9. $\times$ 1 6 8 2
          5 8

10. $\times$ 2 5 8 4
          1 9

11. $\times$ 1 3 0 2
          1 1

12. $\times$ 2 4 8 6
          2 8

1.
$$\times \begin{array}{r} 3\ 1\ 5\ 4 \\ 2\ 4 \end{array}$$

2.
$$\times \begin{array}{r} 1\ 8\ 1\ 6 \\ 3\ 2 \end{array}$$

3.
$$\times \begin{array}{r} 1\ 3\ 6\ 0 \\ 4\ 2 \end{array}$$

4.
$$\times \begin{array}{r} 3\ 6\ 7\ 8 \\ 2\ 9 \end{array}$$

5.
$$\times \begin{array}{r} 2\ 6\ 3\ 8 \\ 3\ 9 \end{array}$$

6.
$$\times \begin{array}{r} 3\ 4\ 4\ 6 \\ 1\ 5 \end{array}$$

7.
$$\times \begin{array}{r} 3\ 5\ 7\ 0 \\ 1\ 9 \end{array}$$

8.
$$\times \begin{array}{r} 1\ 9\ 6\ 5 \\ 4\ 5 \end{array}$$

9.
$$\times \begin{array}{r} 3\ 0\ 9\ 6 \\ 1\ 8 \end{array}$$

10.
$$\times \begin{array}{r} 2\ 4\ 8\ 9 \\ 1\ 7 \end{array}$$

11.
$$\times \begin{array}{r} 2\ 0\ 3\ 6 \\ 2\ 6 \end{array}$$

12.
$$\times \begin{array}{r} 2\ 5\ 0\ 3 \\ 2\ 9 \end{array}$$

1.
$$\begin{array}{r} 3\ 7\ 8\ 2 \\ \times\quad 2\ 7 \\ \hline \end{array}$$

2.
$$\begin{array}{r} 3\ 1\ 6\ 6 \\ \times\quad 1\ 6 \\ \hline \end{array}$$

3.
$$\begin{array}{r} 1\ 7\ 0\ 7 \\ \times\quad 1\ 7 \\ \hline \end{array}$$

4.
$$\begin{array}{r} 1\ 8\ 7\ 8 \\ \times\quad 3\ 9 \\ \hline \end{array}$$

5.
$$\begin{array}{r} 3\ 2\ 1\ 7 \\ \times\quad 2\ 4 \\ \hline \end{array}$$

6.
$$\begin{array}{r} 2\ 7\ 9\ 5 \\ \times\quad 2\ 6 \\ \hline \end{array}$$

7.
$$\begin{array}{r} 2\ 1\ 1\ 4 \\ \times\quad 4\ 2 \\ \hline \end{array}$$

8.
$$\begin{array}{r} 1\ 4\ 9\ 6 \\ \times\quad 6\ 7 \\ \hline \end{array}$$

9.
$$\begin{array}{r} 3\ 7\ 1\ 1 \\ \times\quad 1\ 4 \\ \hline \end{array}$$

10.
$$\begin{array}{r} 3\ 8\ 6\ 3 \\ \times\quad 1\ 6 \\ \hline \end{array}$$

11.
$$\begin{array}{r} 2\ 4\ 9\ 1 \\ \times\quad 3\ 4 \\ \hline \end{array}$$

12.
$$\begin{array}{r} 2\ 4\ 5\ 1 \\ \times\quad 1\ 6 \\ \hline \end{array}$$

1. $\times$ 1 8 1 1
       2 2

2. $\times$ 2 8 4 6
       3 5

3. $\times$ 2 7 9 4
       2 5

4. $\times$ 1 9 9 6
       3 1

5. $\times$ 2 6 2 3
       1 6

6. $\times$ 2 3 7 4
       3 4

7. $\times$ 3 2 1 9
       2 7

8. $\times$ 3 3 1 1
       2 1

9. $\times$ 3 3 3 8
       1 3

10. $\times$ 2 7 4 2
       2 9

11. $\times$ 3 1 3 9
       2 5

12. $\times$ 3 6 4 9
       1 6

1.
$$\begin{array}{r} 2\,8\,7\,0 \\ \times 5\,4 \\ \hline \end{array}$$

2.
$$\begin{array}{r} 3\,9\,2\,1 \\ \times 3\,5 \\ \hline \end{array}$$

3.
$$\begin{array}{r} 2\,1\,5\,6 \\ \times 6\,4 \\ \hline \end{array}$$

4.
$$\begin{array}{r} 3\,4\,5\,0 \\ \times 8\,2 \\ \hline \end{array}$$

5.
$$\begin{array}{r} 4\,3\,2\,7 \\ \times 6\,1 \\ \hline \end{array}$$

6.
$$\begin{array}{r} 3\,3\,2\,6 \\ \times 5\,6 \\ \hline \end{array}$$

7.
$$\begin{array}{r} 2\,7\,8\,9 \\ \times 6\,2 \\ \hline \end{array}$$

8.
$$\begin{array}{r} 3\,5\,5\,5 \\ \times 4\,2 \\ \hline \end{array}$$

9.
$$\begin{array}{r} 2\,6\,7\,3 \\ \times 7\,9 \\ \hline \end{array}$$

10.
$$\begin{array}{r} 2\,7\,5\,4 \\ \times 6\,9 \\ \hline \end{array}$$

11.
$$\begin{array}{r} 4\,0\,6\,6 \\ \times 6\,5 \\ \hline \end{array}$$

12.
$$\begin{array}{r} 3\,4\,6\,8 \\ \times 4\,8 \\ \hline \end{array}$$

1.
$$\begin{array}{r} 4\ 5\ 3\ 4 \\ \times \quad 4\ 6 \\ \hline \end{array}$$

2.
$$\begin{array}{r} 4\ 9\ 0\ 4 \\ \times \quad 8\ 4 \\ \hline \end{array}$$

3.
$$\begin{array}{r} 4\ 1\ 2\ 4 \\ \times \quad 4\ 9 \\ \hline \end{array}$$

4.
$$\begin{array}{r} 2\ 8\ 5\ 0 \\ \times \quad 5\ 4 \\ \hline \end{array}$$

5.
$$\begin{array}{r} 3\ 0\ 8\ 8 \\ \times \quad 4\ 6 \\ \hline \end{array}$$

6.
$$\begin{array}{r} 2\ 6\ 9\ 8 \\ \times \quad 8\ 2 \\ \hline \end{array}$$

7.
$$\begin{array}{r} 3\ 9\ 4\ 5 \\ \times \quad 3\ 8 \\ \hline \end{array}$$

8.
$$\begin{array}{r} 2\ 1\ 2\ 5 \\ \times \quad 8\ 5 \\ \hline \end{array}$$

9.
$$\begin{array}{r} 3\ 2\ 2\ 4 \\ \times \quad 4\ 3 \\ \hline \end{array}$$

10.
$$\begin{array}{r} 3\ 5\ 6\ 2 \\ \times \quad 5\ 7 \\ \hline \end{array}$$

11.
$$\begin{array}{r} 4\ 4\ 1\ 9 \\ \times \quad 2\ 9 \\ \hline \end{array}$$

12.
$$\begin{array}{r} 3\ 4\ 9\ 2 \\ \times \quad 6\ 8 \\ \hline \end{array}$$

1.
$$\begin{array}{r} 2\ 2\ 4\ 8 \\ \times\quad 6\ 3 \\ \hline \end{array}$$

2.
$$\begin{array}{r} 4\ 2\ 2\ 7 \\ \times\quad 3\ 8 \\ \hline \end{array}$$

3.
$$\begin{array}{r} 5\ 7\ 4\ 3 \\ \times\quad 2\ 3 \\ \hline \end{array}$$

4.
$$\begin{array}{r} 5\ 1\ 4\ 5 \\ \times\quad 2\ 7 \\ \hline \end{array}$$

5.
$$\begin{array}{r} 5\ 3\ 3\ 0 \\ \times\quad 3\ 2 \\ \hline \end{array}$$

6.
$$\begin{array}{r} 3\ 8\ 2\ 5 \\ \times\quad 7\ 1 \\ \hline \end{array}$$

7.
$$\begin{array}{r} 3\ 6\ 8\ 9 \\ \times\quad 3\ 8 \\ \hline \end{array}$$

8.
$$\begin{array}{r} 2\ 7\ 4\ 1 \\ \times\quad 5\ 7 \\ \hline \end{array}$$

9.
$$\begin{array}{r} 5\ 5\ 5\ 6 \\ \times\quad 3\ 7 \\ \hline \end{array}$$

10.
$$\begin{array}{r} 3\ 9\ 5\ 8 \\ \times\quad 7\ 1 \\ \hline \end{array}$$

11.
$$\begin{array}{r} 5\ 1\ 7\ 1 \\ \times\quad 4\ 3 \\ \hline \end{array}$$

12.
$$\begin{array}{r} 4\ 6\ 1\ 6 \\ \times\quad 8\ 9 \\ \hline \end{array}$$

1.
$$\begin{array}{r} 1\,9\,8\,6 \\ \times \quad 9\,6 \\ \hline \end{array}$$

2.
$$\begin{array}{r} 3\,6\,6\,4 \\ \times \quad 4\,7 \\ \hline \end{array}$$

3.
$$\begin{array}{r} 5\,2\,7\,2 \\ \times \quad 3\,8 \\ \hline \end{array}$$

4.
$$\begin{array}{r} 4\,1\,6\,4 \\ \times \quad 4\,6 \\ \hline \end{array}$$

5.
$$\begin{array}{r} 4\,9\,9\,1 \\ \times \quad 5\,8 \\ \hline \end{array}$$

6.
$$\begin{array}{r} 4\,4\,3\,9 \\ \times \quad 5\,6 \\ \hline \end{array}$$

7.
$$\begin{array}{r} 2\,2\,4\,6 \\ \times \quad 7\,5 \\ \hline \end{array}$$

8.
$$\begin{array}{r} 5\,0\,1\,3 \\ \times \quad 7\,2 \\ \hline \end{array}$$

9.
$$\begin{array}{r} 5\,7\,5\,4 \\ \times \quad 4\,4 \\ \hline \end{array}$$

10.
$$\begin{array}{r} 3\,8\,3\,2 \\ \times \quad 4\,7 \\ \hline \end{array}$$

11.
$$\begin{array}{r} 4\,2\,8\,3 \\ \times \quad 3\,6 \\ \hline \end{array}$$

12.
$$\begin{array}{r} 4\,1\,9\,7 \\ \times \quad 5\,1 \\ \hline \end{array}$$

1.
$$\begin{array}{r} 3\,6\,6\,1 \\ \times6\,6 \\ \hline \end{array}$$

2.
$$\begin{array}{r} 2\,6\,9\,7 \\ \times9\,2 \\ \hline \end{array}$$

3.
$$\begin{array}{r} 3\,2\,5\,6 \\ \times6\,1 \\ \hline \end{array}$$

4.
$$\begin{array}{r} 5\,6\,3\,7 \\ \times5\,3 \\ \hline \end{array}$$

5.
$$\begin{array}{r} 5\,6\,0\,4 \\ \times3\,7 \\ \hline \end{array}$$

6.
$$\begin{array}{r} 4\,5\,8\,2 \\ \times4\,9 \\ \hline \end{array}$$

7.
$$\begin{array}{r} 2\,3\,9\,6 \\ \times5\,3 \\ \hline \end{array}$$

8.
$$\begin{array}{r} 6\,7\,4\,7 \\ \times6\,3 \\ \hline \end{array}$$

9.
$$\begin{array}{r} 4\,6\,7\,6 \\ \times5\,4 \\ \hline \end{array}$$

10.
$$\begin{array}{r} 5\,2\,4\,2 \\ \times4\,1 \\ \hline \end{array}$$

11.
$$\begin{array}{r} 4\,7\,6\,5 \\ \times6\,9 \\ \hline \end{array}$$

12.
$$\begin{array}{r} 5\,8\,5\,9 \\ \times7\,8 \\ \hline \end{array}$$

Name: **Lesson 2-1 Dividing by two digit numbers with remainders**

1. $22\overline{)1694}$

2. $15\overline{)2716}$

3. $16\overline{)2313}$

5. $17\overline{)2006}$

6. $29\overline{)2642}$

7. $34\overline{)3177}$

9. $13\overline{)2079}$

10. $18\overline{)3724}$

11. $22\overline{)2941}$

13. $33\overline{)3129}$

14. $19\overline{)2115}$

15. $17\overline{)3387}$

Name: **Lesson 2-2 Dividing by two digit numbers with remainders**

1. $11\overline{)1306}$

2. $15\overline{)2405}$

3. $19\overline{)2557}$

5. $21\overline{)3493}$

6. $27\overline{)3599}$

7. $31\overline{)3333}$

9. $17\overline{)2147}$

10. $19\overline{)2169}$

11. $26\overline{)3478}$

13. $25\overline{)3181}$

14. $13\overline{)1850}$

15. $17\overline{)2352}$

1. $16 \overline{)1831}$ 2. $14 \overline{)2617}$ 3. $12 \overline{)2019}$

5. $13 \overline{)1605}$ 6. $19 \overline{)3664}$ 7. $19 \overline{)2459}$

9. $25 \overline{)2764}$ 10. $22 \overline{)2515}$ 11. $23 \overline{)3292}$

13. $27 \overline{)3748}$ 14. $24 \overline{)3394}$ 15. $17 \overline{)2901}$

Name: **Lesson 2-4 Dividing by two digit numbers with remainders**

---

1. $14\,\overline{)2638}$
2. $29\,\overline{)4212}$
3. $22\,\overline{)2431}$

5. $27\,\overline{)3187}$
6. $17\,\overline{)2770}$
7. $17\,\overline{)3188}$

9. $31\,\overline{)4201}$
10. $32\,\overline{)4239}$
11. $34\,\overline{)4235}$

13. $23\,\overline{)4800}$
14. $25\,\overline{)4014}$
15. $21\,\overline{)3400}$

Name: **Lesson 2-5 Dividing by two digit numbers with remainders**

1. $12\,\overline{)3467}$ 2. $21\,\overline{)3926}$ 3. $26\,\overline{)4491}$

5. $15\,\overline{)3264}$ 6. $18\,\overline{)2430}$ 7. $17\,\overline{)2108}$

9. $14\,\overline{)2337}$ 10. $19\,\overline{)3490}$ 11. $21\,\overline{)2241}$

13. $17\,\overline{)2094}$ 14. $39\,\overline{)4021}$ 15. $15\,\overline{)2706}$

Name: **Lesson 2-6 Dividing by two digit numbers with remainders**

1. $25\overline{)4043}$

2. $42\overline{)4931}$

3. $36\overline{)4060}$

5. $19\overline{)5567}$

6. $12\overline{)3661}$

7. $28\overline{)3781}$

9. $11\overline{)2188}$

10. $22\overline{)3491}$

11. $24\overline{)5503}$

13. $23\overline{)2619}$

14. $32\overline{)3901}$

15. $14\overline{)4393}$

Name:

Lesson 2-7 Dividing by two digit numbers with remainders

1. $38\,\overline{)5529}$

2. $44\,\overline{)4779}$

3. $25\,\overline{)3042}$

5. $16\,\overline{)4773}$

6. $12\,\overline{)2116}$

7. $32\,\overline{)4812}$

9. $13\,\overline{)3558}$

10. $27\,\overline{)4544}$

11. $24\,\overline{)5797}$

13. $32\,\overline{)4511}$

14. $18\,\overline{)5959}$

15. $42\,\overline{)5661}$

1. $12 \overline{)4179}$ 　　　2. $16 \overline{)4923}$ 　　　3. $35 \overline{)5540}$

5. $15 \overline{)2227}$ 　　　6. $23 \overline{)5278}$ 　　　7. $27 \overline{)4334}$

9. $39 \overline{)4051}$ 　　　10. $18 \overline{)3577}$ 　　　11. $41 \overline{)6192}$

13. $19 \overline{)5943}$ 　　　14. $22 \overline{)6935}$ 　　　15. $29 \overline{)3456}$

Name: **Lesson 2-9 Dividing by two digit numbers with remainders**

1. $55\,\overline{)6237}$

2. $37\,\overline{)6805}$

3. $12\,\overline{)5929}$

5. $33\,\overline{)5070}$

6. $28\,\overline{)6763}$

7. $23\,\overline{)4824}$

9. $26\,\overline{)3319}$

10. $26\,\overline{)3723}$

11. $32\,\overline{)5737}$

13. $11\,\overline{)2896}$

14. $42\,\overline{)5565}$

15. $44\,\overline{)6611}$

1. $13\overline{)5562}$ 2. $12\overline{)3189}$ 3. $31\overline{)5254}$

5. $25\overline{)3528}$ 6. $46\overline{)5635}$ 7. $19\overline{)4920}$

9. $35\overline{)6174}$ 10. $16\overline{)3857}$ 11. $39\overline{)5280}$

13. $41\overline{)5542}$ 14. $23\overline{)5472}$ 15. $28\overline{)6555}$

Change each fraction to simplest form.

1. $\dfrac{2}{4}$ =   2. $\dfrac{4}{12}$ =   3. $\dfrac{3}{15}$ =

4. $\dfrac{2}{6}$ =   5. $\dfrac{8}{12}$ =   6. $\dfrac{4}{12}$ =

7. $\dfrac{3}{9}$ =   8. $\dfrac{4}{6}$ =   9. $\dfrac{6}{9}$ =

10. $\dfrac{2}{8}$ =   11. $\dfrac{2}{12}$ =   12. $\dfrac{10}{12}$ =

13. $\dfrac{4}{8}$ =   14. $\dfrac{6}{9}$ =   15. $\dfrac{5}{15}$ =

16. $\dfrac{6}{8}$ =   17. $\dfrac{6}{12}$ =   18. $\dfrac{12}{18}$ =

19. $\dfrac{2}{2}$ =   20. $\dfrac{3}{9}$ =   21. $\dfrac{8}{12}$ =

22. $\dfrac{6}{12}$ =   23. $\dfrac{10}{12}$ =   24. $\dfrac{9}{12}$ =

25. $\dfrac{3}{6}$ =   26. $\dfrac{9}{9}$ =   27. $\dfrac{15}{18}$ =

Change each fraction to simplest form.

1. $\dfrac{5}{10} =$ 2. $\dfrac{9}{12} =$ 3. $\dfrac{12}{24} =$

4. $\dfrac{4}{6} =$ 5. $\dfrac{6}{10} =$ 6. $\dfrac{2}{18} =$

7. $\dfrac{8}{12} =$ 8. $\dfrac{4}{12} =$ 9. $\dfrac{15}{18} =$

10. $\dfrac{6}{10} =$ 11. $\dfrac{4}{10} =$ 12. $\dfrac{4}{24} =$

13. $\dfrac{6}{8} =$ 14. $\dfrac{5}{15} =$ 15. $\dfrac{10}{18} =$

16. $\dfrac{10}{15} =$ 17. $\dfrac{3}{18} =$ 18. $\dfrac{17}{34} =$

19. $\dfrac{3}{12} =$ 20. $\dfrac{2}{6} =$ 21. $\dfrac{6}{24} =$

22. $\dfrac{2}{10} =$ 23. $\dfrac{6}{12} =$ 24. $\dfrac{7}{28} =$

25. $\dfrac{12}{15} =$ 26. $\dfrac{12}{20} =$ 27. $\dfrac{15}{30} =$

Change each fraction to simplest form.

1. $\dfrac{3}{12}$ =

2. $\dfrac{4}{8}$ =

3. $\dfrac{8}{12}$ =

4. $\dfrac{6}{8}$ =

5. $\dfrac{2}{6}$ =

6. $\dfrac{2}{8}$ =

7. $\dfrac{4}{6}$ =

8. $\dfrac{4}{10}$ =

9. $\dfrac{8}{10}$ =

10. $\dfrac{6}{10}$ =

11. $\dfrac{4}{12}$ =

12. $\dfrac{12}{15}$ =

13. $\dfrac{7}{14}$ =

14. $\dfrac{5}{15}$ =

15. $\dfrac{9}{18}$ =

16. $\dfrac{10}{15}$ =

17. $\dfrac{15}{18}$ =

18. $\dfrac{18}{24}$ =

19. $\dfrac{6}{12}$ =

20. $\dfrac{16}{24}$ =

21. $\dfrac{22}{24}$ =

22. $\dfrac{12}{18}$ =

23. $\dfrac{20}{24}$ =

24. $\dfrac{10}{12}$ =

25. $\dfrac{10}{24}$ =

26. $\dfrac{10}{25}$ =

27. $\dfrac{20}{25}$ =

Change each fraction to simplest form.

1. $\dfrac{4}{10} =$ 2. $\dfrac{6}{12} =$ 3. $\dfrac{3}{15} =$

4. $\dfrac{10}{12} =$ 5. $\dfrac{6}{10} =$ 6. $\dfrac{8}{12} =$

7. $\dfrac{10}{15} =$ 8. $\dfrac{5}{15} =$ 9. $\dfrac{8}{10} =$

10. $\dfrac{6}{18} =$ 11. $\dfrac{8}{18} =$ 12. $\dfrac{10}{18} =$

13. $\dfrac{12}{20} =$ 14. $\dfrac{2}{20} =$ 15. $\dfrac{5}{20} =$

16. $\dfrac{16}{20} =$ 17. $\dfrac{18}{20} =$ 18. $\dfrac{4}{24} =$

19. $\dfrac{18}{24} =$ 20. $\dfrac{20}{24} =$ 21. $\dfrac{22}{24} =$

22. $\dfrac{15}{30} =$ 23. $\dfrac{6}{30} =$ 24. $\dfrac{8}{30} =$

25. $\dfrac{4}{32} =$ 26. $\dfrac{6}{32} =$ 27. $\dfrac{20}{32} =$

Fill in the blank with $<$, $>$, or $=$ to make each statement true.

1. $\dfrac{1}{2}$ —— 1 2. $\dfrac{3}{3}$ —— 1 3. $\dfrac{1}{2}$ —— $\dfrac{6}{12}$

4. $\dfrac{1}{4}$ —— $\dfrac{2}{3}$ 5. $\dfrac{5}{4}$ —— $\dfrac{2}{3}$ 6. $\dfrac{4}{3}$ —— $\dfrac{2}{3}$

7. $\dfrac{1}{3}$ —— $\dfrac{1}{6}$ 8. $\dfrac{2}{6}$ —— $\dfrac{1}{3}$ 9. $\dfrac{8}{9}$ —— $\dfrac{4}{6}$

10. $\dfrac{2}{5}$ —— $\dfrac{3}{5}$ 11. $\dfrac{3}{7}$ —— $\dfrac{2}{3}$ 12. $\dfrac{10}{3}$ —— 1

13. 1 —— $\dfrac{3}{4}$ 14. $\dfrac{6}{10}$ —— $\dfrac{3}{5}$ 15. $\dfrac{25}{20}$ —— $\dfrac{5}{4}$

16. $\dfrac{5}{6}$ —— $\dfrac{1}{2}$ 17. $\dfrac{6}{5}$ —— 1 18. $\dfrac{11}{33}$ —— $\dfrac{2}{6}$

19. $\dfrac{1}{3}$ —— $\dfrac{1}{5}$ 20. $\dfrac{12}{4}$ —— 1 21. $\dfrac{2}{3}$ —— $\dfrac{9}{12}$

22. $\dfrac{2}{5}$ —— $\dfrac{1}{2}$ 23. $\dfrac{5}{8}$ —— $\dfrac{1}{2}$ 24. $\dfrac{5}{6}$ —— $\dfrac{5}{7}$

25. $\dfrac{2}{7}$ —— $\dfrac{3}{7}$ 26. $\dfrac{9}{5}$ —— $\dfrac{6}{7}$ 27. $\dfrac{21}{28}$ —— $\dfrac{12}{16}$

Fill in the blank with $<$, $>$, or $=$ to make each statement true.

1. $\dfrac{2}{3}$ ___ $\dfrac{5}{9}$ 2. $\dfrac{2}{2}$ ___ $\dfrac{4}{3}$ 3. $\dfrac{1}{4}$ ___ $\dfrac{1}{8}$

4. $\dfrac{1}{2}$ ___ $\dfrac{3}{4}$ 5. $\dfrac{1}{3}$ ___ $\dfrac{1}{2}$ 6. $\dfrac{3}{4}$ ___ $\dfrac{9}{12}$

7. $\dfrac{3}{4}$ ___ $\dfrac{1}{3}$ 8. 1 ___ $\dfrac{4}{3}$ 9. $\dfrac{3}{3}$ ___ $\dfrac{7}{7}$

10. $\dfrac{2}{5}$ ___ $\dfrac{2}{6}$ 11. $\dfrac{4}{7}$ ___ $\dfrac{1}{2}$ 12. 1 ___ $\dfrac{7}{6}$

13. $\dfrac{3}{7}$ ___ $\dfrac{4}{8}$ 14. $\dfrac{2}{8}$ ___ $\dfrac{1}{3}$ 15. $\dfrac{5}{15}$ ___ $\dfrac{3}{9}$

16. $\dfrac{5}{8}$ ___ $\dfrac{1}{2}$ 17. $\dfrac{10}{12}$ ___ $\dfrac{20}{24}$ 18. $\dfrac{8}{12}$ ___ $\dfrac{10}{20}$

19. $\dfrac{6}{12}$ ___ $\dfrac{2}{3}$ 20. $\dfrac{8}{7}$ ___ $\dfrac{5}{5}$ 21. $\dfrac{9}{12}$ ___ $\dfrac{4}{6}$

22. $\dfrac{9}{18}$ ___ $\dfrac{11}{22}$ 23. $\dfrac{24}{8}$ ___ 1 24. $\dfrac{21}{28}$ ___ $\dfrac{27}{36}$

25. $\dfrac{15}{30}$ ___ $\dfrac{16}{32}$ 26. $\dfrac{7}{21}$ ___ $\dfrac{6}{18}$ 27. $\dfrac{6}{5}$ ___ $\dfrac{18}{15}$

Fill in the blank with $<$, $>$, or $=$ to make each statement true.

1. $\dfrac{4}{5}$ —— $\dfrac{3}{4}$ 2. $\dfrac{2}{6}$ —— $\dfrac{1}{4}$ 3. $\dfrac{5}{5}$ —— $\dfrac{5}{6}$

4. $\dfrac{1}{6}$ —— $\dfrac{3}{3}$ 5. $\dfrac{2}{5}$ —— $\dfrac{6}{15}$ 6. $\dfrac{8}{9}$ —— $\dfrac{16}{18}$

7. $\dfrac{2}{7}$ —— $\dfrac{1}{6}$ 8. $\dfrac{3}{7}$ —— $\dfrac{2}{3}$ 9. $\dfrac{4}{5}$ —— $\dfrac{2}{3}$

10. $\dfrac{14}{16}$ —— $\dfrac{5}{6}$ 11. $\dfrac{4}{6}$ —— $\dfrac{10}{15}$ 12. $\dfrac{7}{10}$ —— $\dfrac{1}{2}$

13. $\dfrac{17}{5}$ —— $\dfrac{15}{7}$ 14. $\dfrac{7}{21}$ —— $\dfrac{2}{3}$ 15. $\dfrac{9}{15}$ —— $\dfrac{5}{10}$

16. $\dfrac{11}{21}$ —— $\dfrac{13}{21}$ 17. $\dfrac{4}{3}$ —— $\dfrac{16}{12}$ 18. $\dfrac{9}{5}$ —— $\dfrac{19}{6}$

19. $\dfrac{13}{6}$ —— $\dfrac{11}{6}$ 20. $\dfrac{7}{2}$ —— $\dfrac{19}{7}$ 21. $\dfrac{24}{4}$ —— $\dfrac{36}{9}$

22. $\dfrac{9}{18}$ —— $\dfrac{17}{34}$ 23. $\dfrac{28}{35}$ —— $\dfrac{16}{20}$ 24. $\dfrac{8}{9}$ —— $\dfrac{24}{27}$

25. $\dfrac{12}{30}$ —— $\dfrac{4}{10}$ 26. $\dfrac{4}{10}$ —— $\dfrac{9}{15}$ 27. $\dfrac{7}{21}$ —— $\dfrac{16}{48}$

Name: Lesson 3-8   Fractions

Find equivalent fractions.

1. $\dfrac{1}{2} = \dfrac{2}{4} = \dfrac{3}{6} = \dfrac{5}{10} = \dfrac{6}{12} = \dfrac{9}{18}$

2. $\dfrac{1}{3} = \dfrac{\ }{6} = \dfrac{\ }{12} = \dfrac{5}{\ } = \dfrac{\ }{21} = \dfrac{10}{\ }$

3. $\dfrac{1}{5} = \dfrac{3}{\ } = \dfrac{\ }{20} = \dfrac{6}{\ } = \dfrac{\ }{35} = \dfrac{10}{\ }$

4. $\dfrac{2}{3} = \dfrac{4}{\ } = \dfrac{6}{\ } = \dfrac{\ }{15} = \dfrac{12}{\ } = \dfrac{\ }{24}$

5. $\dfrac{2}{5} = \dfrac{6}{\ } = \dfrac{\ }{25} = \dfrac{\ }{30} = \dfrac{16}{\ } = \dfrac{\ }{55}$

6. $\dfrac{1}{6} = \dfrac{2}{\ } = \dfrac{4}{\ } = \dfrac{\ }{42} = \dfrac{\ }{48} = \dfrac{9}{\ }$

7. $\dfrac{3}{4} = \dfrac{\ }{8} = \dfrac{\ }{12} = \dfrac{\ }{20} = \dfrac{21}{\ } = \dfrac{27}{\ }$

8. $\dfrac{3}{5} = \dfrac{\ }{15} = \dfrac{12}{\ } = \dfrac{15}{\ } = \dfrac{18}{\ } = \dfrac{\ }{55}$

9. $\dfrac{5}{6} = \dfrac{10}{\ } = \dfrac{20}{\ } = \dfrac{\ }{30} = \dfrac{\ }{36} = \dfrac{45}{\ }$

Find equivalent fractions.

1. $\dfrac{2}{3} = \dfrac{}{6} = \dfrac{8}{} = \dfrac{10}{} = \dfrac{}{21} = \dfrac{}{24}$

2. $\dfrac{3}{4} = \dfrac{}{12} = \dfrac{}{16} = \dfrac{18}{} = \dfrac{21}{} = \dfrac{}{36}$

3. $\dfrac{3}{5} = \dfrac{9}{} = \dfrac{}{20} = \dfrac{15}{} = \dfrac{}{35} = \dfrac{24}{}$

4. $\dfrac{2}{7} = \dfrac{}{14} = \dfrac{6}{} = \dfrac{10}{} = \dfrac{}{42} = \dfrac{}{56}$

5. $\dfrac{4}{5} = \dfrac{}{10} = \dfrac{}{20} = \dfrac{20}{} = \dfrac{24}{} = \dfrac{}{50}$

6. $\dfrac{3}{7} = \dfrac{9}{} = \dfrac{15}{} = \dfrac{}{42} = \dfrac{}{49} = \dfrac{27}{}$

7. $\dfrac{4}{9} = \dfrac{}{27} = \dfrac{}{45} = \dfrac{28}{} = \dfrac{32}{} = \dfrac{}{81}$

8. $\dfrac{5}{7} = \dfrac{10}{} = \dfrac{25}{} = \dfrac{}{42} = \dfrac{35}{} = \dfrac{}{56}$

9. $\dfrac{5}{9} = \dfrac{}{36} = \dfrac{}{45} = \dfrac{30}{} = \dfrac{35}{} = \dfrac{}{72}$

Find equivalent fractions.

1. $\dfrac{5}{7} = \dfrac{10}{} = \dfrac{}{21} = \dfrac{35}{} = \dfrac{}{56} = \dfrac{45}{}$

2. $\dfrac{2}{9} = \dfrac{}{18} = \dfrac{6}{} = \dfrac{}{54} = \dfrac{14}{} = \dfrac{}{81}$

3. $\dfrac{5}{6} = \dfrac{}{18} = \dfrac{}{24} = \dfrac{30}{} = \dfrac{35}{} = \dfrac{}{66}$

4. $\dfrac{6}{7} = \dfrac{18}{} = \dfrac{24}{} = \dfrac{}{49} = \dfrac{}{56} = \dfrac{}{84}$

5. $\dfrac{4}{9} = \dfrac{}{45} = \dfrac{}{54} = \dfrac{36}{} = \dfrac{40}{} = \dfrac{}{108}$

6. $\dfrac{2}{7} = \dfrac{10}{} = \dfrac{}{42} = \dfrac{16}{} = \dfrac{}{63} = \dfrac{}{91}$

7. $\dfrac{3}{10} = \dfrac{9}{} = \dfrac{}{40} = \dfrac{}{70} = \dfrac{24}{} = \dfrac{45}{}$

8. $\dfrac{7}{9} = \dfrac{21}{} = \dfrac{}{36} = \dfrac{49}{} = \dfrac{}{72} = \dfrac{98}{}$

9. $\dfrac{9}{11} = \dfrac{}{22} = \dfrac{36}{} = \dfrac{}{55} = \dfrac{81}{} = \dfrac{}{121}$

Add fractions and write the answers in simplest form.

1. $\dfrac{1}{6} + \dfrac{1}{6} =$

2. $\dfrac{1}{6} + \dfrac{2}{6} =$

3. $\dfrac{1}{4} + \dfrac{1}{4} =$

4. $\dfrac{2}{4} + \dfrac{6}{4} =$

5. $\dfrac{2}{10} + \dfrac{3}{10} =$

6. $\dfrac{1}{10} + \dfrac{5}{10} =$

7. $\dfrac{2}{8} + \dfrac{4}{8} =$

8. $\dfrac{3}{5} + \dfrac{2}{5} =$

9. $\dfrac{1}{5} + \dfrac{2}{5} =$

10. $\dfrac{1}{8} + \dfrac{3}{8} =$

11. $\dfrac{1}{9} + \dfrac{2}{9} =$

12. $\dfrac{3}{12} + \dfrac{3}{12} =$

13. $\dfrac{2}{12} + \dfrac{1}{12} =$

14. $\dfrac{2}{9} + \dfrac{4}{9} =$

15. $\dfrac{3}{10} + \dfrac{5}{10} =$

16. $\dfrac{5}{10} + \dfrac{1}{10} =$

17. $\dfrac{3}{12} + \dfrac{2}{12} =$

18. $\dfrac{4}{12} + \dfrac{6}{12} =$

Add fractions and write the answers in simplest form.

1. $\dfrac{3}{4} + \dfrac{3}{4} =$

2. $\dfrac{1}{8} + \dfrac{3}{8} =$

3. $\dfrac{2}{5} + \dfrac{3}{5} =$

4. $\dfrac{1}{4} + \dfrac{1}{4} =$

5. $\dfrac{3}{8} + \dfrac{3}{8} =$

6. $\dfrac{3}{10} + \dfrac{2}{10} =$

7. $\dfrac{1}{10} + \dfrac{1}{10} =$

8. $\dfrac{4}{9} + \dfrac{2}{9} =$

9. $\dfrac{2}{9} + \dfrac{1}{9} =$

10. $\dfrac{1}{12} + \dfrac{1}{12} =$

11. $\dfrac{2}{12} + \dfrac{2}{12} =$

12. $\dfrac{7}{12} + \dfrac{9}{12} =$

13. $\dfrac{7}{12} + \dfrac{3}{12} =$

14. $\dfrac{4}{15} + \dfrac{6}{15} =$

15. $\dfrac{3}{15} + \dfrac{2}{15} =$

16. $\dfrac{8}{15} + \dfrac{7}{15} =$

17. $\dfrac{1}{14} + \dfrac{1}{14} =$

18. $\dfrac{5}{14} + \dfrac{2}{14} =$

Name: _____

Lesson 4-3 Adding fractions with the same denominators

Add fractions and write the answers in simplest form.

1. $\dfrac{1}{2} + \dfrac{3}{2} =$

2. $\dfrac{1}{4} + \dfrac{7}{4} =$

3. $\dfrac{2}{4} + \dfrac{5}{4} =$

4. $\dfrac{1}{6} + \dfrac{2}{6} =$

5. $\dfrac{3}{6} + \dfrac{1}{6} =$

6. $\dfrac{5}{8} + \dfrac{7}{8} =$

7. $\dfrac{6}{8} + \dfrac{4}{8} =$

8. $\dfrac{7}{9} + \dfrac{8}{9} =$

9. $\dfrac{7}{9} + \dfrac{5}{9} =$

10. $\dfrac{6}{12} + \dfrac{8}{12} =$

11. $\dfrac{4}{12} + \dfrac{4}{12} =$

12. $\dfrac{10}{12} + \dfrac{8}{12} =$

13. $\dfrac{5}{10} + \dfrac{7}{10} =$

14. $\dfrac{8}{10} + \dfrac{12}{10} =$

15. $\dfrac{6}{15} + \dfrac{9}{15} =$

16. $\dfrac{6}{15} + \dfrac{6}{15} =$

17. $\dfrac{10}{12} + \dfrac{10}{12} =$

18. $\dfrac{12}{8} + \dfrac{6}{8} =$

Add fractions and write the answers in simplest form.

1. $\dfrac{1}{2} + \dfrac{1}{2} =$

2. $1\dfrac{1}{3} + \dfrac{1}{3} =$

3. $1\dfrac{1}{3} + 1\dfrac{2}{3} =$

4. $\dfrac{1}{4} + 2\dfrac{3}{4} =$

5. $2\dfrac{1}{4} + \dfrac{1}{4} =$

6. $3\dfrac{1}{6} + 1\dfrac{1}{6} =$

7. $2\dfrac{1}{6} + 2\dfrac{2}{6} =$

8. $1\dfrac{1}{8} + \dfrac{3}{8} =$

9. $\dfrac{1}{8} + 4\dfrac{3}{8} =$

10. $4\dfrac{2}{8} + 3\dfrac{2}{8} =$

11. $2\dfrac{3}{8} + 3\dfrac{3}{8} =$

12. $2\dfrac{1}{9} + 4\dfrac{5}{9} =$

13. $3\dfrac{2}{9} + 3\dfrac{1}{9} =$

14. $4\dfrac{3}{10} + \dfrac{1}{10} =$

15. $5\dfrac{5}{10} + \dfrac{5}{10} =$

16. $4\dfrac{1}{12} + 4\dfrac{1}{12} =$

17. $5\dfrac{2}{12} + 4\dfrac{1}{12} =$

18. $\dfrac{3}{15} + 3\dfrac{2}{15} =$

Add fractions and write the answers in simplest form.

1. $2\dfrac{3}{6} + 1\dfrac{1}{6} =$

2. $3\dfrac{1}{8} + 2\dfrac{3}{8} =$

3. $1\dfrac{1}{8} + \dfrac{1}{8} =$

4. $2\dfrac{1}{10} + 3\dfrac{3}{10} =$

5. $3\dfrac{3}{10} + 2\dfrac{2}{10} =$

6. $5\dfrac{2}{8} + \dfrac{4}{8} =$

7. $5\dfrac{1}{12} + 4\dfrac{1}{12} =$

8. $4\dfrac{1}{12} + 3\dfrac{2}{12} =$

9. $\dfrac{3}{12} + 5\dfrac{3}{12} =$

10. $6\dfrac{5}{12} + 4\dfrac{3}{12} =$

11. $1\dfrac{2}{15} + 2\dfrac{1}{15} =$

12. $\dfrac{2}{15} + 1\dfrac{3}{15} =$

13. $4\dfrac{3}{18} + 2\dfrac{3}{18} =$

14. $5\dfrac{5}{18} + 6\dfrac{4}{18} =$

15. $\dfrac{3}{20} + 5\dfrac{7}{20} =$

16. $2\dfrac{5}{16} + \dfrac{3}{16} =$

17. $7\dfrac{1}{16} + 3\dfrac{1}{16} =$

18. $4\dfrac{3}{20} + 7\dfrac{2}{20} =$

Subtract fractions and write the answers in simplest form.

1. $\dfrac{2}{2} - \dfrac{1}{2} =$

2. $\dfrac{3}{4} - \dfrac{2}{4} =$

3. $\dfrac{3}{4} - \dfrac{1}{4} =$

4. $\dfrac{5}{8} - \dfrac{1}{8} =$

5. $\dfrac{4}{5} - \dfrac{2}{5} =$

6. $\dfrac{7}{8} - \dfrac{3}{8} =$

7. $\dfrac{3}{6} - \dfrac{1}{6} =$

8. $\dfrac{8}{10} - \dfrac{4}{10} =$

9. $\dfrac{9}{10} - \dfrac{4}{10} =$

10. $\dfrac{7}{12} - \dfrac{3}{12} =$

11. $\dfrac{5}{10} - \dfrac{3}{10} =$

12. $\dfrac{11}{12} - \dfrac{1}{12} =$

13. $\dfrac{12}{12} - \dfrac{2}{12} =$

14. $\dfrac{10}{15} - \dfrac{3}{15} =$

15. $\dfrac{8}{12} - \dfrac{5}{12} =$

16. $\dfrac{13}{18} - \dfrac{4}{18} =$

17. $\dfrac{12}{18} - \dfrac{10}{18} =$

18. $\dfrac{14}{15} - \dfrac{4}{15} =$

Subtract fractions and write the answers in simplest form.

1. $\dfrac{4}{6} - \dfrac{2}{6} =$

2. $\dfrac{8}{9} - \dfrac{2}{9} =$

3. $\dfrac{4}{7} - \dfrac{2}{7} =$

4. $\dfrac{5}{8} - \dfrac{2}{8} =$

5. $\dfrac{7}{9} - \dfrac{1}{9} =$

6. $\dfrac{7}{10} - \dfrac{2}{10} =$

7. $\dfrac{11}{10} - \dfrac{3}{10} =$

8. $\dfrac{7}{12} - \dfrac{2}{12} =$

9. $\dfrac{15}{12} - \dfrac{6}{12} =$

10. $\dfrac{9}{15} - \dfrac{5}{15} =$

11. $\dfrac{11}{15} - \dfrac{2}{15} =$

12. $\dfrac{13}{14} - \dfrac{6}{14} =$

13. $\dfrac{9}{14} - \dfrac{3}{14} =$

14. $\dfrac{15}{16} - \dfrac{3}{16} =$

15. $\dfrac{10}{11} - \dfrac{3}{11} =$

16. $\dfrac{10}{13} - \dfrac{3}{13} =$

17. $\dfrac{14}{20} - \dfrac{8}{20} =$

18. $\dfrac{18}{20} - \dfrac{4}{20} =$

Subtract fractions and write the answers in simplest form.

1. $\dfrac{6}{7} - \dfrac{2}{7} =$

2. $\dfrac{3}{8} - \dfrac{2}{8} =$

3. $\dfrac{7}{8} - \dfrac{5}{8} =$

4. $\dfrac{7}{9} - \dfrac{1}{9} =$

5. $\dfrac{5}{9} - \dfrac{2}{9} =$

6. $\dfrac{9}{10} - \dfrac{3}{10} =$

7. $\dfrac{7}{10} - \dfrac{2}{10} =$

8. $\dfrac{11}{12} - \dfrac{8}{12} =$

9. $\dfrac{11}{12} - \dfrac{2}{12} =$

10. $\dfrac{9}{14} - \dfrac{2}{14} =$

11. $\dfrac{8}{13} - \dfrac{1}{13} =$

12. $\dfrac{14}{15} - \dfrac{2}{15} =$

13. $\dfrac{6}{11} - \dfrac{3}{11} =$

14. $\dfrac{11}{18} - \dfrac{5}{18} =$

15. $\dfrac{17}{18} - \dfrac{2}{18} =$

16. $\dfrac{7}{17} - \dfrac{4}{17} =$

17. $\dfrac{13}{16} - \dfrac{3}{16} =$

18. $\dfrac{18}{20} - \dfrac{3}{20} =$

Subtract fractions and write the answers in simplest form.

1. $1\dfrac{2}{3} - \dfrac{1}{3} =$

2. $1\dfrac{2}{5} - \dfrac{1}{5} =$

3. $3\dfrac{3}{4} - 1\dfrac{1}{4} =$

4. $3\dfrac{6}{8} - 1\dfrac{4}{8} =$

5. $2\dfrac{3}{6} - 2\dfrac{1}{6} =$

6. $4\dfrac{7}{9} - 2\dfrac{1}{9} =$

7. $5\dfrac{2}{5} - \dfrac{1}{5} =$

8. $5\dfrac{7}{10} - 4\dfrac{5}{10} =$

9. $4\dfrac{4}{8} - 2\dfrac{2}{8} =$

10. $3\dfrac{7}{12} - 3\dfrac{3}{12} =$

11. $6\dfrac{6}{10} - \dfrac{2}{10} =$

12. $7\dfrac{11}{12} - 2\dfrac{1}{12} =$

13. $7\dfrac{9}{12} - 2\dfrac{3}{12} =$

14. $6\dfrac{5}{13} - \dfrac{2}{13} =$

15. $4\dfrac{9}{15} - 1\dfrac{3}{15} =$

16. $4\dfrac{7}{15} - 1\dfrac{4}{15} =$

17. $6\dfrac{12}{18} - 3\dfrac{2}{18} =$

18. $7\dfrac{10}{18} - 3\dfrac{6}{18} =$

Name: _____

Subtract fractions and write the answers in simplest form.

1. $3\dfrac{5}{6} - \dfrac{2}{6} =$

2. $4\dfrac{4}{5} - \dfrac{2}{5} =$

3. $2\dfrac{5}{8} - 1\dfrac{4}{8} =$

4. $3\dfrac{3}{4} - 1\dfrac{1}{4} =$

5. $4\dfrac{7}{9} - 2\dfrac{4}{9} =$

6. $5\dfrac{5}{6} - 2\dfrac{1}{6} =$

7. $3\dfrac{8}{10} - 1\dfrac{4}{10} =$

8. $7\dfrac{7}{11} - 4\dfrac{3}{11} =$

9. $6\dfrac{10}{12} - 3\dfrac{5}{12} =$

10. $5\dfrac{8}{12} - 3\dfrac{4}{12} =$

11. $5\dfrac{8}{13} - 2\dfrac{3}{18} =$

12. $4\dfrac{9}{13} - \dfrac{3}{13} =$

13. $7\dfrac{8}{15} - 4\dfrac{5}{15} =$

14. $6\dfrac{14}{15} - 5\dfrac{5}{15} =$

15. $6\dfrac{9}{16} - 3\dfrac{5}{16} =$

16. $4\dfrac{11}{16} - \dfrac{1}{16} =$

17. $5\dfrac{12}{18} - \dfrac{9}{18} =$

18. $6\dfrac{13}{18} - 4\dfrac{3}{18} =$

Multiply fractions and write the answers in simplest form.

1. $2 \times \dfrac{3}{4} =$

2. $\dfrac{3}{4} \times 8 =$

3. $2 \times \dfrac{3}{2} =$

4. $\dfrac{2}{5} \times 5 =$

5. $3 \times \dfrac{5}{6} =$

6. $\dfrac{3}{7} \times 7 =$

7. $4 \times \dfrac{2}{6} =$

8. $\dfrac{3}{8} \times 4 =$

9. $5 \times \dfrac{7}{10} =$

10. $\dfrac{5}{8} \times 12 =$

11. $5 \times \dfrac{3}{10} =$

12. $\dfrac{7}{10} \times 5 =$

13. $4 \times \dfrac{5}{6} =$

14. $\dfrac{4}{12} \times 3 =$

15. $4 \times \dfrac{7}{12} =$

16. $\dfrac{5}{9} \times 3 =$

17. $6 \times \dfrac{9}{12} =$

18. $\dfrac{7}{15} \times 5 =$

Multiply fractions and write the answers in simplest form.

1. $\dfrac{1}{2} \times \dfrac{2}{3} =$

2. $\dfrac{1}{3} \times \dfrac{3}{4} =$

3. $\dfrac{1}{3} \times \dfrac{2}{5} =$

4. $\dfrac{1}{2} \times \dfrac{1}{3} =$

5. $\dfrac{1}{2} \times \dfrac{4}{5} =$

6. $\dfrac{1}{3} \times \dfrac{3}{5} =$

7. $\dfrac{1}{4} \times \dfrac{4}{7} =$

8. $\dfrac{3}{7} \times \dfrac{2}{3} =$

9. $\dfrac{2}{3} \times \dfrac{3}{4} =$

10. $\dfrac{2}{3} \times \dfrac{3}{6} =$

11. $\dfrac{3}{5} \times \dfrac{5}{6} =$

12. $\dfrac{4}{5} \times \dfrac{5}{8} =$

13. $\dfrac{7}{8} \times \dfrac{4}{7} =$

14. $\dfrac{2}{3} \times \dfrac{4}{7} =$

15. $\dfrac{3}{4} \times \dfrac{3}{5} =$

16. $\dfrac{4}{9} \times \dfrac{3}{8} =$

17. $\dfrac{2}{6} \times \dfrac{9}{12} =$

18. $\dfrac{3}{12} \times \dfrac{8}{9} =$

Multiply fractions and write the answers in simplest form.

1. $\dfrac{1}{3} \times \dfrac{3}{6} =$

2. $\dfrac{5}{4} \times \dfrac{1}{7} =$

3. $\dfrac{3}{8} \times \dfrac{3}{5} =$

4. $\dfrac{5}{2} \times \dfrac{6}{5} =$

5. $\dfrac{3}{5} \times \dfrac{8}{9} =$

6. $\dfrac{3}{5} \times \dfrac{5}{3} =$

7. $\dfrac{1}{2} \times \dfrac{6}{8} =$

8. $\dfrac{6}{7} \times \dfrac{7}{9} =$

9. $\dfrac{2}{7} \times \dfrac{3}{4} =$

10. $\dfrac{2}{6} \times \dfrac{2}{4} =$

11. $\dfrac{4}{9} \times \dfrac{1}{3} =$

12. $\dfrac{1}{9} \times \dfrac{3}{4} =$

13. $\dfrac{2}{5} \times \dfrac{4}{6} =$

14. $\dfrac{2}{6} \times \dfrac{8}{4} =$

15. $\dfrac{3}{2} \times \dfrac{4}{6} =$

16. $\dfrac{3}{7} \times \dfrac{5}{3} =$

17. $\dfrac{1}{7} \times \dfrac{1}{8} =$

18. $\dfrac{3}{8} \times \dfrac{6}{12} =$

Multiply fractions and write the answers in simplest form.

1. $\dfrac{3}{4} \times \dfrac{2}{5} =$

2. $\dfrac{2}{3} \times \dfrac{2}{10} =$

3. $\dfrac{2}{9} \times \dfrac{9}{5} =$

4. $\dfrac{6}{11} \times \dfrac{10}{12} =$

5. $\dfrac{7}{4} \times \dfrac{12}{5} =$

6. $\dfrac{2}{6} \times \dfrac{7}{2} =$

7. $\dfrac{11}{5} \times \dfrac{10}{7} =$

8. $\dfrac{8}{3} \times \dfrac{1}{6} =$

9. $\dfrac{5}{9} \times \dfrac{12}{5} =$

10. $\dfrac{7}{3} \times \dfrac{3}{4} =$

11. $\dfrac{9}{8} \times \dfrac{12}{3} =$

12. $\dfrac{9}{11} \times \dfrac{7}{2} =$

13. $\dfrac{5}{7} \times \dfrac{12}{4} =$

14. $\dfrac{9}{10} \times \dfrac{3}{11} =$

15. $\dfrac{8}{3} \times \dfrac{3}{8} =$

16. $\dfrac{6}{12} \times \dfrac{8}{7} =$

17. $\dfrac{6}{4} \times \dfrac{8}{9} =$

18. $\dfrac{3}{12} \times \dfrac{9}{6} =$

Multiply fractions and write the answers in simplest form.

1. $\dfrac{4}{5} \times \dfrac{10}{6} =$

2. $\dfrac{7}{12} \times \dfrac{4}{7} =$

3. $\dfrac{3}{7} \times \dfrac{14}{15} =$

4. $\dfrac{6}{2} \times \dfrac{4}{13} =$

5. $\dfrac{12}{7} \times \dfrac{14}{7} =$

6. $\dfrac{2}{6} \times \dfrac{11}{6} =$

7. $\dfrac{5}{3} \times \dfrac{12}{8} =$

8. $\dfrac{2}{14} \times \dfrac{10}{6} =$

9. $\dfrac{7}{9} \times \dfrac{3}{14} =$

10. $\dfrac{14}{8} \times \dfrac{12}{7} =$

11. $\dfrac{6}{5} \times \dfrac{2}{12} =$

12. $\dfrac{9}{15} \times \dfrac{12}{8} =$

13. $\dfrac{6}{15} \times \dfrac{10}{11} =$

14. $\dfrac{13}{5} \times \dfrac{10}{8} =$

15. $\dfrac{3}{11} \times \dfrac{10}{6} =$

16. $\dfrac{8}{12} \times \dfrac{6}{4} =$

17. $\dfrac{5}{8} \times \dfrac{5}{15} =$

18. $\dfrac{7}{2} \times \dfrac{8}{9} =$

Multiply fractions and write the answers in simplest form.

1. $\dfrac{7}{6} \times \dfrac{6}{3} =$ 2. $\dfrac{14}{13} \times \dfrac{11}{7} =$

3. $\dfrac{11}{8} \times \dfrac{7}{22} =$ 4. $\dfrac{10}{4} \times \dfrac{14}{8} =$

5. $\dfrac{12}{17} \times \dfrac{7}{16} =$ 6. $\dfrac{2}{11} \times \dfrac{22}{7} =$

7. $\dfrac{2}{15} \times \dfrac{9}{13} =$ 8. $\dfrac{3}{5} \times \dfrac{10}{9} =$

9. $\dfrac{10}{9} \times \dfrac{9}{10} =$ 10. $\dfrac{14}{16} \times \dfrac{4}{7} =$

11. $\dfrac{4}{18} \times \dfrac{9}{10} =$ 12. $\dfrac{12}{5} \times \dfrac{2}{16} =$

13. $\dfrac{16}{17} \times \dfrac{8}{12} =$ 14. $\dfrac{5}{16} \times \dfrac{6}{15} =$

15. $\dfrac{3}{8} \times \dfrac{4}{9} =$ 16. $\dfrac{6}{2} \times \dfrac{4}{14} =$

17. $\dfrac{4}{16} \times \dfrac{8}{11} =$ 18. $\dfrac{4}{17} \times \dfrac{34}{16} =$

Multiply fractions and write the answers in simplest form.

1. $\dfrac{4}{5} \times \dfrac{6}{14} =$ 2. $\dfrac{12}{13} \times \dfrac{13}{24} =$

3. $\dfrac{9}{10} \times \dfrac{14}{21} =$ 4. $\dfrac{11}{3} \times \dfrac{9}{33} =$

5. $\dfrac{17}{14} \times \dfrac{7}{34} =$ 6. $\dfrac{16}{3} \times \dfrac{15}{24} =$

7. $\dfrac{13}{24} \times \dfrac{12}{7} =$ 8. $\dfrac{12}{21} \times \dfrac{14}{9} =$

9. $\dfrac{22}{19} \times \dfrac{38}{11} =$ 10. $\dfrac{8}{22} \times \dfrac{2}{11} =$

11. $\dfrac{8}{20} \times \dfrac{13}{8} =$ 12. $\dfrac{3}{4} \times \dfrac{11}{9} =$

13. $\dfrac{17}{9} \times \dfrac{12}{17} =$ 14. $\dfrac{16}{13} \times \dfrac{13}{17} =$

15. $\dfrac{5}{15} \times \dfrac{9}{10} =$ 16. $\dfrac{11}{17} \times \dfrac{19}{11} =$

17. $\dfrac{13}{16} \times \dfrac{14}{7} =$ 18. $\dfrac{2}{8} \times \dfrac{10}{11} =$

Multiply fractions and write the answers in simplest form.

1. $\dfrac{3}{20} \times \dfrac{10}{18} =$

2. $\dfrac{4}{11} \times \dfrac{33}{12} =$

3. $\dfrac{12}{5} \times \dfrac{15}{24} =$

4. $\dfrac{18}{12} \times \dfrac{17}{6} =$

5. $\dfrac{9}{30} \times \dfrac{18}{3} =$

6. $\dfrac{15}{9} \times \dfrac{18}{30} =$

7. $\dfrac{22}{3} \times \dfrac{9}{33} =$

8. $\dfrac{4}{22} \times \dfrac{33}{2} =$

9. $\dfrac{12}{9} \times \dfrac{15}{3} =$

10. $\dfrac{23}{4} \times \dfrac{2}{23} =$

11. $\dfrac{12}{11} \times \dfrac{17}{24} =$

12. $\dfrac{18}{28} \times \dfrac{7}{9} =$

13. $\dfrac{10}{19} \times \dfrac{38}{40} =$

14. $\dfrac{21}{19} \times \dfrac{19}{28} =$

15. $\dfrac{10}{3} \times \dfrac{2}{4} =$

16. $\dfrac{16}{29} \times \dfrac{2}{8} =$

17. $\dfrac{13}{17} \times \dfrac{34}{26} =$

18. $\dfrac{5}{10} \times \dfrac{30}{15} =$

Multiply fractions and write the answers in simplest form.

1. $\dfrac{13}{24} \times \dfrac{12}{5} =$

2. $\dfrac{12}{15} \times \dfrac{21}{18} =$

3. $\dfrac{18}{12} \times \dfrac{9}{10} =$

4. $\dfrac{16}{5} \times \dfrac{15}{12} =$

5. $\dfrac{9}{25} \times \dfrac{10}{3} =$

6. $\dfrac{2}{12} \times \dfrac{3}{7} =$

7. $\dfrac{7}{6} \times \dfrac{3}{21} =$

8. $\dfrac{4}{9} \times \dfrac{27}{12} =$

9. $\dfrac{17}{24} \times \dfrac{8}{5} =$

10. $\dfrac{5}{3} \times \dfrac{8}{10} =$

11. $\dfrac{11}{18} \times \dfrac{9}{33} =$

12. $\dfrac{4}{22} \times \dfrac{11}{12} =$

13. $\dfrac{13}{20} \times \dfrac{10}{39} =$

14. $\dfrac{5}{27} \times \dfrac{3}{20} =$

15. $\dfrac{5}{10} \times \dfrac{20}{8} =$

16. $\dfrac{9}{15} \times \dfrac{5}{6} =$

17. $\dfrac{27}{18} \times \dfrac{9}{18} =$

18. $\dfrac{12}{21} \times \dfrac{7}{18} =$

Multiply fractions and write the answers in simplest form.

1. $\dfrac{22}{27} \times \dfrac{9}{10} =$

2. $\dfrac{26}{4} \times \dfrac{14}{13} =$

3. $\dfrac{16}{25} \times \dfrac{15}{8} =$

4. $\dfrac{10}{28} \times \dfrac{14}{5} =$

5. $\dfrac{23}{14} \times \dfrac{21}{23} =$

6. $\dfrac{25}{6} \times \dfrac{24}{15} =$

7. $\dfrac{7}{22} \times \dfrac{33}{14} =$

8. $\dfrac{11}{28} \times \dfrac{21}{33} =$

9. $\dfrac{13}{2} \times \dfrac{8}{39} =$

10. $\dfrac{4}{28} \times \dfrac{14}{2} =$

11. $\dfrac{16}{17} \times \dfrac{34}{22} =$

12. $\dfrac{13}{4} \times \dfrac{12}{39} =$

13. $\dfrac{14}{16} \times \dfrac{8}{7} =$

14. $\dfrac{26}{17} \times \dfrac{8}{39} =$

15. $\dfrac{16}{17} \times \dfrac{17}{18} =$

16. $\dfrac{3}{19} \times \dfrac{57}{6} =$

17. $\dfrac{27}{36} \times \dfrac{12}{18} =$

18. $\dfrac{8}{21} \times \dfrac{42}{24} =$

Divide fractions and write the answers in simplest form.

1. $\dfrac{2}{3} \div 3 =$

2. $2 \div \dfrac{2}{3} =$

3. $\dfrac{1}{2} \div 2 =$

4. $3 \div \dfrac{1}{3} =$

5. $\dfrac{3}{4} \div 3 =$

6. $4 \div \dfrac{8}{5} =$

7. $\dfrac{2}{5} \div 4 =$

8. $6 \div \dfrac{12}{7} =$

9. $\dfrac{1}{6} \div 3 =$

10. $5 \div \dfrac{10}{15} =$

11. $\dfrac{2}{4} \div 3 =$

12. $4 \div \dfrac{8}{7} =$

13. $\dfrac{3}{5} \div 6 =$

14. $8 \div \dfrac{12}{5} =$

15. $\dfrac{4}{7} \div 6 =$

16. $9 \div \dfrac{6}{4} =$

17. $\dfrac{4}{5} \div 12 =$

18. $10 \div \dfrac{20}{3} =$

Divide fractions and write the answers in simplest form.

1. $\dfrac{1}{8} \div \dfrac{3}{8} =$ 2. $\dfrac{3}{2} \div \dfrac{7}{2} =$

3. $\dfrac{5}{7} \div \dfrac{5}{6} =$ 4. $\dfrac{7}{5} \div \dfrac{3}{5} =$

5. $\dfrac{3}{4} \div \dfrac{2}{3} =$ 6. $\dfrac{2}{7} \div \dfrac{7}{6} =$

7. $\dfrac{2}{6} \div \dfrac{8}{3} =$ 8. $\dfrac{7}{6} \div \dfrac{3}{5} =$

9. $\dfrac{2}{5} \div \dfrac{10}{3} =$ 10. $\dfrac{6}{4} \div \dfrac{3}{5} =$

11. $\dfrac{3}{8} \div \dfrac{6}{4} =$ 12. $\dfrac{8}{3} \div \dfrac{8}{5} =$

13. $\dfrac{2}{7} \div \dfrac{6}{2} =$ 14. $\dfrac{3}{5} \div \dfrac{8}{6} =$

15. $\dfrac{4}{3} \div \dfrac{8}{7} =$ 16. $\dfrac{6}{5} \div \dfrac{8}{9} =$

17. $\dfrac{4}{5} \div \dfrac{4}{6} =$ 18. $\dfrac{5}{2} \div \dfrac{5}{3} =$

Divide fractions and write the answers in simplest form.

1. $\dfrac{3}{5} \div \dfrac{4}{3} =$ 2. $\dfrac{4}{6} \div \dfrac{4}{8} =$

3. $\dfrac{9}{12} \div \dfrac{10}{4} =$ 4. $\dfrac{1}{6} \div \dfrac{11}{12} =$

5. $\dfrac{2}{9} \div \dfrac{12}{5} =$ 6. $\dfrac{10}{12} \div \dfrac{4}{7} =$

7. $\dfrac{1}{2} \div \dfrac{6}{7} =$ 8. $\dfrac{6}{7} \div \dfrac{7}{3} =$

9. $\dfrac{9}{6} \div \dfrac{3}{5} =$ 10. $\dfrac{10}{7} \div \dfrac{3}{6} =$

11. $\dfrac{4}{10} \div \dfrac{12}{9} =$ 12. $\dfrac{2}{10} \div \dfrac{3}{10} =$

13. $\dfrac{4}{12} \div \dfrac{10}{3} =$ 14. $\dfrac{3}{9} \div \dfrac{10}{12} =$

15. $\dfrac{2}{10} \div \dfrac{10}{12} =$ 16. $\dfrac{11}{4} \div \dfrac{9}{6} =$

17. $\dfrac{11}{7} \div \dfrac{8}{14} =$ 18. $\dfrac{3}{7} \div \dfrac{4}{10} =$

68

Divide fractions and write the answers in simplest form.

1. $\dfrac{4}{12} \div \dfrac{4}{5} =$ 2. $\dfrac{5}{11} \div \dfrac{11}{4} =$

3. $\dfrac{9}{8} \div \dfrac{2}{7} =$ 4. $\dfrac{10}{9} \div \dfrac{12}{9} =$

5. $\dfrac{6}{2} \div \dfrac{7}{8} =$ 6. $\dfrac{11}{6} \div \dfrac{9}{10} =$

7. $\dfrac{5}{8} \div \dfrac{4}{5} =$ 8. $\dfrac{2}{7} \div \dfrac{10}{3} =$

9. $\dfrac{6}{9} \div \dfrac{11}{8} =$ 10. $\dfrac{2}{8} \div \dfrac{5}{4} =$

11. $\dfrac{6}{5} \div \dfrac{3}{5} =$ 12. $\dfrac{8}{4} \div \dfrac{5}{9} =$

13. $\dfrac{10}{7} \div \dfrac{8}{7} =$ 14. $\dfrac{6}{10} \div \dfrac{7}{3} =$

15. $\dfrac{5}{9} \div \dfrac{9}{4} =$ 16. $\dfrac{4}{10} \div \dfrac{2}{5} =$

17. $\dfrac{5}{11} \div \dfrac{2}{5} =$ 18. $\dfrac{3}{11} \div \dfrac{12}{3} =$

Divide fractions and write the answers in simplest form.

1. $\dfrac{1}{7} \div \dfrac{3}{4} =$ 2. $\dfrac{3}{5} \div \dfrac{9}{12} =$

3. $\dfrac{4}{6} \div \dfrac{6}{7} =$ 4. $\dfrac{5}{11} \div \dfrac{7}{11} =$

5. $\dfrac{12}{11} \div \dfrac{5}{10} =$ 6. $\dfrac{2}{7} \div \dfrac{8}{6} =$

7. $\dfrac{3}{12} \div \dfrac{2}{12} =$ 8. $\dfrac{3}{5} \div \dfrac{5}{7} =$

9. $\dfrac{2}{6} \div \dfrac{4}{7} =$ 10. $\dfrac{5}{12} \div \dfrac{10}{6} =$

11. $\dfrac{1}{4} \div \dfrac{8}{11} =$ 12. $\dfrac{3}{8} \div \dfrac{3}{4} =$

13. $\dfrac{4}{10} \div \dfrac{8}{6} =$ 14. $\dfrac{12}{7} \div \dfrac{9}{11} =$

15. $\dfrac{3}{8} \div \dfrac{6}{5} =$ 16. $\dfrac{2}{4} \div \dfrac{3}{6} =$

17. $\dfrac{1}{3} \div \dfrac{6}{8} =$ 18. $\dfrac{10}{7} \div \dfrac{4}{3} =$

Divide fractions and write the answers in simplest form.

1. $\dfrac{13}{12} \div \dfrac{9}{13} =$

2. $\dfrac{5}{8} \div \dfrac{7}{11} =$

3. $\dfrac{8}{12} \div \dfrac{8}{5} =$

4. $\dfrac{6}{10} \div \dfrac{3}{12} =$

5. $\dfrac{14}{11} \div \dfrac{7}{5} =$

6. $\dfrac{13}{11} \div \dfrac{11}{3} =$

7. $\dfrac{9}{10} \div \dfrac{5}{8} =$

8. $\dfrac{15}{2} \div \dfrac{6}{5} =$

9. $\dfrac{1}{12} \div \dfrac{4}{14} =$

10. $\dfrac{2}{9} \div \dfrac{12}{7} =$

11. $\dfrac{7}{3} \div \dfrac{14}{5} =$

12. $\dfrac{10}{13} \div \dfrac{8}{5} =$

13. $\dfrac{14}{10} \div \dfrac{14}{10} =$

14. $\dfrac{12}{15} \div \dfrac{4}{10} =$

15. $\dfrac{15}{10} \div \dfrac{12}{5} =$

16. $\dfrac{4}{3} \div \dfrac{12}{7} =$

17. $\dfrac{10}{11} \div \dfrac{14}{15} =$

18. $\dfrac{14}{2} \div \dfrac{7}{13} =$

Divide fractions and write the answers in simplest form.

1. $\dfrac{5}{6} \div \dfrac{12}{15} =$

2. $\dfrac{10}{11} \div \dfrac{6}{7} =$

3. $\dfrac{6}{3} \div \dfrac{3}{9} =$

4. $\dfrac{1}{7} \div \dfrac{6}{3} =$

5. $\dfrac{7}{15} \div \dfrac{6}{5} =$

6. $\dfrac{15}{10} \div \dfrac{2}{3} =$

7. $\dfrac{3}{13} \div \dfrac{26}{14} =$

8. $\dfrac{13}{7} \div \dfrac{39}{5} =$

9. $\dfrac{11}{10} \div \dfrac{6}{14} =$

10. $\dfrac{15}{14} \div \dfrac{15}{7} =$

11. $\dfrac{7}{2} \div \dfrac{8}{9} =$

12. $\dfrac{14}{12} \div \dfrac{7}{3} =$

13. $\dfrac{14}{3} \div \dfrac{9}{10} =$

14. $\dfrac{8}{12} \div \dfrac{11}{3} =$

15. $\dfrac{9}{5} \div \dfrac{4}{12} =$

16. $\dfrac{4}{12} \div \dfrac{5}{14} =$

17. $\dfrac{12}{11} \div \dfrac{9}{6} =$

18. $\dfrac{7}{8} \div \dfrac{7}{9} =$

Divide fractions and write the answers in simplest form.

1. $\dfrac{9}{10} \div \dfrac{6}{8} =$ 2. $\dfrac{13}{14} \div \dfrac{26}{28} =$

3. $\dfrac{10}{16} \div \dfrac{12}{7} =$ 4. $\dfrac{9}{20} \div \dfrac{12}{30} =$

5. $\dfrac{5}{14} \div \dfrac{8}{7} =$ 6. $\dfrac{8}{9} \div \dfrac{24}{18} =$

7. $\dfrac{11}{6} \div \dfrac{10}{22} =$ 8. $\dfrac{9}{17} \div \dfrac{36}{34} =$

9. $\dfrac{14}{3} \div \dfrac{28}{5} =$ 10. $\dfrac{7}{6} \div \dfrac{8}{6} =$

11. $\dfrac{2}{11} \div \dfrac{3}{22} =$ 12. $\dfrac{14}{6} \div \dfrac{18}{5} =$

13. $\dfrac{8}{5} \div \dfrac{12}{15} =$ 14. $\dfrac{7}{17} \div \dfrac{5}{51} =$

15. $\dfrac{3}{13} \div \dfrac{9}{39} =$ 16. $\dfrac{6}{7} \div \dfrac{12}{7} =$

17. $\dfrac{12}{10} \div \dfrac{12}{5} =$ 18. $\dfrac{16}{19} \div \dfrac{8}{19} =$

Divide fractions and write the answers in simplest form.

1. $\dfrac{18}{9} \div \dfrac{2}{13} =$

2. $\dfrac{18}{5} \div \dfrac{6}{35} =$

3. $\dfrac{17}{7} \div \dfrac{17}{11} =$

4. $\dfrac{11}{16} \div \dfrac{33}{24} =$

5. $\dfrac{4}{14} \div \dfrac{18}{7} =$

6. $\dfrac{6}{9} \div \dfrac{4}{3} =$

7. $\dfrac{11}{18} \div \dfrac{22}{12} =$

8. $\dfrac{14}{3} \div \dfrac{8}{5} =$

9. $\dfrac{4}{15} \div \dfrac{18}{7} =$

10. $\dfrac{4}{19} \div \dfrac{12}{38} =$

11. $\dfrac{8}{13} \div \dfrac{6}{26} =$

12. $\dfrac{16}{14} \div \dfrac{4}{8} =$

13. $\dfrac{14}{18} \div \dfrac{7}{6} =$

14. $\dfrac{7}{4} \div \dfrac{4}{6} =$

15. $\dfrac{20}{6} \div \dfrac{5}{12} =$

16. $\dfrac{18}{11} \div \dfrac{2}{22} =$

17. $\dfrac{16}{17} \div \dfrac{4}{34} =$

18. $\dfrac{13}{8} \div \dfrac{13}{2} =$

Divide fractions and write the answers in simplest form.

1. $\dfrac{6}{18} \div \dfrac{12}{3} =$

2. $\dfrac{8}{20} \div \dfrac{7}{15} =$

3. $\dfrac{16}{10} \div \dfrac{8}{5} =$

4. $\dfrac{5}{8} \div \dfrac{7}{8} =$

5. $\dfrac{7}{4} \div \dfrac{8}{12} =$

6. $\dfrac{7}{5} \div \dfrac{6}{15} =$

7. $\dfrac{3}{13} \div \dfrac{9}{5} =$

8. $\dfrac{16}{15} \div \dfrac{11}{15} =$

9. $\dfrac{7}{11} \div \dfrac{14}{4} =$

10. $\dfrac{15}{17} \div \dfrac{5}{6} =$

11. $\dfrac{4}{10} \div \dfrac{2}{11} =$

12. $\dfrac{3}{2} \div \dfrac{4}{7} =$

13. $\dfrac{9}{5} \div \dfrac{11}{10} =$

14. $\dfrac{19}{9} \div \dfrac{38}{6} =$

15. $\dfrac{4}{12} \div \dfrac{8}{17} =$

16. $\dfrac{13}{8} \div \dfrac{39}{12} =$

17. $\dfrac{2}{10} \div \dfrac{11}{9} =$

18. $\dfrac{5}{15} \div \dfrac{15}{30} =$

Change the fractions to decimals and round to the nearest thousandth if necessary.

1. $\dfrac{1}{4} =$

2. $\dfrac{1}{5} =$

3. $\dfrac{3}{5} =$

4. $\dfrac{2}{5} =$

5. $\dfrac{2}{4} =$

6. $\dfrac{6}{8} =$

7. $\dfrac{3}{6} =$

8. $\dfrac{4}{8} =$

9. $\dfrac{3}{4} =$

10. $\dfrac{2}{8} =$

11. $\dfrac{1}{10} =$

12. $\dfrac{5}{8} =$

13. $\dfrac{2}{10} =$

14. $\dfrac{3}{8} =$

15. $\dfrac{4}{5} =$

16. $\dfrac{6}{10} =$

17. $\dfrac{3}{10} =$

18. $\dfrac{7}{8} =$

19. $\dfrac{3}{12} =$

20. $\dfrac{5}{10} =$

21. $\dfrac{4}{10} =$

22. $\dfrac{4}{20} =$

23. $\dfrac{6}{12} =$

24. $\dfrac{9}{12} =$

25. $\dfrac{6}{16} =$

26. $\dfrac{6}{15} =$

27. $\dfrac{4}{16} =$

Change the fractions to decimals and round to the nearest thousandth if necessary.

1. $\dfrac{1}{4} =$ 2. $\dfrac{1}{2} =$ 3. $\dfrac{1}{5} =$

4. $\dfrac{3}{4} =$ 5. $\dfrac{2}{5} =$ 6. $\dfrac{2}{4} =$

7. $\dfrac{3}{5} =$ 8. $\dfrac{2}{8} =$ 9. $\dfrac{4}{5} =$

10. $\dfrac{1}{8} =$ 11. $\dfrac{4}{20} =$ 12. $\dfrac{3}{8} =$

13. $\dfrac{4}{8} =$ 14. $\dfrac{16}{20} =$ 15. $\dfrac{6}{8} =$

16. $\dfrac{7}{8} =$ 17. $\dfrac{5}{8} =$ 18. $\dfrac{14}{20} =$

19. $\dfrac{2}{20} =$ 20. $\dfrac{9}{12} =$ 21. $\dfrac{5}{20} =$

22. $\dfrac{3}{20} =$ 23. $\dfrac{3}{25} =$ 24. $\dfrac{1}{20} =$

25. $\dfrac{4}{25} =$ 26. $\dfrac{6}{25} =$ 27. $\dfrac{5}{25} =$

Change the fractions to decimals and round to the nearest thousandth if necessary.

1. $\dfrac{3}{2} =$ 2. $\dfrac{3}{4} =$ 3. $\dfrac{5}{4} =$

4. $\dfrac{6}{5} =$ 5. $\dfrac{2}{5} =$ 6. $\dfrac{4}{5} =$

7. $\dfrac{9}{4} =$ 8. $\dfrac{7}{5} =$ 9. $\dfrac{5}{2} =$

10. $\dfrac{8}{5} =$ 11. $\dfrac{6}{4} =$ 12. $\dfrac{8}{4} =$

13. $\dfrac{3}{8} =$ 14. $\dfrac{7}{4} =$ 15. $\dfrac{12}{8} =$

16. $\dfrac{9}{6} =$ 17. $\dfrac{10}{8} =$ 18. $\dfrac{9}{5} =$

19. $\dfrac{12}{8} =$ 20. $\dfrac{12}{10} =$ 21. $\dfrac{9}{8} =$

22. $\dfrac{14}{10} =$ 23. $\dfrac{13}{5} =$ 24. $\dfrac{14}{10} =$

25. $\dfrac{11}{5} =$ 26. $\dfrac{13}{8} =$ 27. $\dfrac{13}{5} =$

Change the fractions to decimals and round to the nearest thousandth if necessary.

1. $\dfrac{7}{5} =$

2. $\dfrac{3}{5} =$

3. $\dfrac{3}{4} =$

4. $\dfrac{1}{4} =$

5. $\dfrac{2}{4} =$

6. $\dfrac{9}{5} =$

7. $\dfrac{3}{20} =$

8. $\dfrac{6}{5} =$

9. $\dfrac{9}{6} =$

10. $\dfrac{7}{4} =$

11. $\dfrac{10}{4} =$

12. $\dfrac{13}{4} =$

13. $\dfrac{4}{25} =$

14. $\dfrac{14}{5} =$

15. $\dfrac{1}{20} =$

16. $\dfrac{3}{10} =$

17. $\dfrac{11}{10} =$

18. $\dfrac{12}{5} =$

19. $\dfrac{9}{4} =$

20. $\dfrac{11}{4} =$

21. $\dfrac{3}{20} =$

22. $\dfrac{5}{25} =$

23. $\dfrac{6}{25} =$

24. $\dfrac{11}{20} =$

25. $\dfrac{3}{50} =$

26. $\dfrac{6}{50} =$

27. $\dfrac{7}{25} =$

Change the fractions to decimals and round to the nearest thousandth if necessary.

1. $\dfrac{6}{5} =$

2. $\dfrac{6}{4} =$

3. $\dfrac{7}{2} =$

4. $\dfrac{9}{4} =$

5. $\dfrac{8}{5} =$

6. $\dfrac{7}{4} =$

7. $\dfrac{9}{2} =$

8. $\dfrac{10}{4} =$

9. $\dfrac{13}{5} =$

10. $\dfrac{16}{10} =$

11. $\dfrac{17}{2} =$

12. $\dfrac{11}{4} =$

13. $\dfrac{13}{2} =$

14. $\dfrac{15}{4} =$

15. $\dfrac{9}{25} =$

16. $\dfrac{15}{6} =$

17. $\dfrac{12}{5} =$

18. $\dfrac{13}{4} =$

19. $\dfrac{6}{20} =$

20. $\dfrac{3}{20} =$

21. $\dfrac{16}{5} =$

22. $\dfrac{6}{25} =$

23. $\dfrac{9}{8} =$

24. $\dfrac{18}{12} =$

25. $\dfrac{17}{8} =$

26. $\dfrac{7}{25} =$

27. $\dfrac{15}{8} =$

Change the fractions to decimals and round to the nearest thousandth if necessary.

1. $\dfrac{9}{4} =$

2. $\dfrac{11}{4} =$

3. $\dfrac{13}{4} =$

4. $\dfrac{7}{2} =$

5. $\dfrac{12}{8} =$

6. $\dfrac{24}{16} =$

7. $\dfrac{11}{5} =$

8. $\dfrac{11}{22} =$

9. $\dfrac{17}{5} =$

10. $\dfrac{7}{14} =$

11. $\dfrac{14}{5} =$

12. $\dfrac{11}{44} =$

13. $\dfrac{13}{52} =$

14. $\dfrac{10}{8} =$

15. $\dfrac{1}{20} =$

16. $\dfrac{13}{25} =$

17. $\dfrac{7}{20} =$

18. $\dfrac{19}{5} =$

19. $\dfrac{11}{8} =$

20. $\dfrac{15}{25} =$

21. $\dfrac{14}{25} =$

22. $\dfrac{2}{25} =$

23. $\dfrac{13}{20} =$

24. $\dfrac{9}{20} =$

25. $\dfrac{11}{25} =$

26. $\dfrac{17}{25} =$

27. $\dfrac{19}{8} =$

Change the fractions to decimals and round to the nearest thousandth if necessary.

1. $\dfrac{13}{4} =$ 2. $\dfrac{4}{5} =$ 3. $\dfrac{17}{10} =$

4. $\dfrac{16}{5} =$ 5. $\dfrac{15}{4} =$ 6. $\dfrac{14}{20} =$

7. $\dfrac{17}{8} =$ 8. $\dfrac{27}{6} =$ 9. $\dfrac{7}{100} =$

10. $\dfrac{19}{20} =$ 11. $\dfrac{17}{5} =$ 12. $\dfrac{17}{4} =$

13. $\dfrac{13}{10} =$ 14. $\dfrac{29}{8} =$ 15. $\dfrac{9}{24} =$

16. $\dfrac{4}{32} =$ 17. $\dfrac{11}{25} =$ 18. $\dfrac{18}{5} =$

19. $\dfrac{17}{20} =$ 20. $\dfrac{7}{20} =$ 21. $\dfrac{7}{50} =$

22. $\dfrac{21}{8} =$ 23. $\dfrac{18}{20} =$ 24. $\dfrac{21}{5} =$

25. $\dfrac{17}{25} =$ 26. $\dfrac{1}{100} =$ 27. $\dfrac{4}{100} =$

Change the fractions to decimals and round to the nearest thousandth if necessary.

1. $\dfrac{1}{20} =$

2. $\dfrac{11}{2} =$

3. $\dfrac{13}{4} =$

4. $\dfrac{15}{2} =$

5. $\dfrac{5}{20} =$

6. $\dfrac{21}{6} =$

7. $\dfrac{29}{8} =$

8. $\dfrac{5}{25} =$

9. $\dfrac{11}{25} =$

10. $\dfrac{22}{100} =$

11. $\dfrac{17}{100} =$

12. $\dfrac{7}{20} =$

13. $\dfrac{2}{25} =$

14. $\dfrac{25}{8} =$

15. $\dfrac{11}{100} =$

16. $\dfrac{37}{100} =$

17. $\dfrac{9}{25} =$

18. $\dfrac{13}{25} =$

19. $\dfrac{21}{2} =$

20. $\dfrac{23}{5} =$

21. $\dfrac{9}{20} =$

22. $\dfrac{37}{5} =$

23. $\dfrac{31}{8} =$

24. $\dfrac{7}{100} =$

25. $\dfrac{18}{25} =$

26. $\dfrac{33}{6} =$

27. $\dfrac{13}{20} =$

Change the fractions to decimals and round to the nearest thousandth if necessary.

1. $\dfrac{2}{5} =$

2. $\dfrac{1}{8} =$

3. $\dfrac{1}{25} =$

4. $\dfrac{11}{4} =$

5. $\dfrac{12}{5} =$

6. $\dfrac{1}{20} =$

7. $\dfrac{11}{8} =$

8. $\dfrac{15}{6} =$

9. $\dfrac{19}{5} =$

10. $\dfrac{24}{16} =$

11. $\dfrac{13}{4} =$

12. $\dfrac{18}{24} =$

13. $\dfrac{17}{2} =$

14. $\dfrac{9}{36} =$

15. $\dfrac{17}{4} =$

16. $\dfrac{29}{4} =$

17. $\dfrac{11}{100} =$

18. $\dfrac{7}{20} =$

19. $\dfrac{8}{40} =$

20. $\dfrac{8}{20} =$

21. $\dfrac{3}{25} =$

22. $\dfrac{19}{25} =$

23. $\dfrac{47}{8} =$

24. $\dfrac{13}{52} =$

25. $\dfrac{42}{50} =$

26. $\dfrac{33}{50} =$

27. $\dfrac{49}{50} =$

Change the fractions to decimals and round to the nearest thousandth if necessary.

1. $\dfrac{1}{3} =$

2. $\dfrac{14}{5} =$

3. $\dfrac{1}{6} =$

4. $\dfrac{4}{5} =$

5. $\dfrac{2}{3} =$

6. $\dfrac{6}{25} =$

7. $\dfrac{19}{4} =$

8. $\dfrac{25}{4} =$

9. $\dfrac{2}{20} =$

10. $\dfrac{42}{5} =$

11. $\dfrac{3}{20} =$

12. $\dfrac{57}{8} =$

13. $\dfrac{9}{100} =$

14. $\dfrac{43}{8} =$

15. $\dfrac{27}{18} =$

16. $\dfrac{4}{3} =$

17. $\dfrac{23}{2} =$

18. $\dfrac{7}{6} =$

19. $\dfrac{22}{25} =$

20. $\dfrac{4}{6} =$

21. $\dfrac{65}{8} =$

22. $\dfrac{17}{20} =$

23. $\dfrac{29}{50} =$

24. $\dfrac{10}{3} =$

25. $\dfrac{45}{100} =$

26. $\dfrac{23}{25} =$

27. $\dfrac{18}{27} =$

Change decimals to fractions and write the answers in simplest form.

1. $0.2 =$

2. $0.25 =$

3. $0.75 =$

4. $0.5 =$

5. $0.6 =$

6. $0.1 =$

7. $0.01 =$

8. $0.05 =$

9. $0.4 =$

10. $0.6 =$

11. $0.9 =$

12. $0.8 =$

13. $0.15 =$

14. $0.35 =$

15. $0.02 =$

16. $0.04 =$

17. $0.65 =$

18. $0.7 =$

19. $0.3 =$

20. $0.85 =$

Name:

Change decimals to fractions and write the answers in simplest form.

1. $0.1 =$

2. $0.3 =$

3. $0.2 =$

4. $0.4 =$

5. $0.35 =$

6. $0.15 =$

7. $0.25 =$

8. $0.09 =$

9. $0.03 =$

10. $0.55 =$

11. $0.45 =$

12. $0.11 =$

13. $0.21 =$

14. $0.33 =$

15. $0.65 =$

16. $0.05 =$

17. $0.75 =$

18. $0.95 =$

19. $0.53 =$

20. $0.07 =$

Change decimals to fractions and write the answers in simplest form.

1. $0.01 =$

2. $0.05 =$

3. $0.8 =$

4. $0.25 =$

5. $0.75 =$

6. $0.16 =$

7. $0.125 =$

8. $0.09 =$

9. $0.88 =$

10. $0.08 =$

11. $0.87 =$

12. $0.99 =$

13. $0.625 =$

14. $0.875 =$

15. $0.24 =$

16. $0.15 =$

17. $0.4 =$

18. $0.375 =$

19. $0.57 =$

20. $0.24 =$

Change decimals to fractions and write the answers in simplest form.

1. $0.05 =$

2. $0.3 =$

3. $1.2 =$

4. $1.1 =$

5. $0.95 =$

6. $0.875 =$

7. $2.75 =$

8. $0.45 =$

9. $0.85 =$

10. $0.125 =$

11. $1.11 =$

12. $1.6 =$

13. $0.65 =$

14. $0.35 =$

15. $0.375 =$

16. $3.25 =$

17. $0.55 =$

18. $1.45 =$

19. $3.65 =$

20. $1.125 =$

Lesson 8-5   Changing decimals to fractions

Change decimals to fractions and write the answers in simplest form.

1. $1.5 =$

2. $2.4 =$

3. $0.38 =$

4. $0.44 =$

5. $1.75 =$

6. $4.25 =$

7. $1.45 =$

8. $0.05 =$

9. $1.375 =$

10. $0.68 =$

11. $0.52 =$

12. $2.625 =$

13. $2.7 =$

14. $3.55 =$

15. $1.98 =$

16. $0.87 =$

17. $3.1 =$

18. $1.23 =$

19. $5.75 =$

20. $4.99 =$

Change decimals to fractions and write the answers in simplest form.

1. $1.12 =$

2. $2.8 =$

3. $4.125 =$

4. $6.25 =$

5. $1.96 =$

6. $1.86 =$

7. $0.76 =$

8. $0.28 =$

9. $1.58 =$

10. $4.55 =$

11. $5.75 =$

12. $3.875 =$

13. $2.625 =$

14. $1.92 =$

15. $1.35 =$

16. $0.64 =$

17. $0.78 =$

18. $3.86 =$

19. $2.18 =$

20. $2.48 =$

Change decimals to fractions and write the answers in simplest form.

1. $2.2 =$ 2. $0.15 =$

3. $1.08 =$ 4. $0.4 =$

5. $0.62 =$ 6. $3.25 =$

7. $5.75 =$ 8. $2.6 =$

9. $4.35 =$ 10. $0.85 =$

11. $0.03 =$ 12. $3.04 =$

13. $5.4 =$ 14. $4.82 =$

15. $0.48 =$ 16. $6.75 =$

17. $2.56 =$ 18. $1.875 =$

19. $2.01 =$ 20. $5.52 =$

Lesson 8-8 Changing decimals to fractions

Change decimals to fractions and write the answers in simplest form.

1. $4.09 =$

2. $0.75 =$

3. $0.45 =$

4. $1.53 =$

5. $1.76 =$

6. $3.74 =$

7. $6.18 =$

8. $6.3 =$

9. $0.81 =$

10. $4.625 =$

11. $4.55 =$

12. $8.01 =$

13. $8.12 =$

14. $8.02 =$

15. $7.2 =$

16. $9.41 =$

17. $4.125 =$

18. $1.48 =$

19. $1.09 =$

20. $3.98 =$

Change decimals to fractions and write the answers in simplest form.

1. $0.35 =$ 2. $0.14 =$

3. $1.68 =$ 4. $2.06 =$

5. $0.97 =$ 6. $5.9 =$

7. $4.82 =$ 8. $6.89 =$

9. $8.25 =$ 10. $7.75 =$

11. $6.92 =$ 12. $1.6 =$

13. $7.625 =$ 14. $2.44 =$

15. $4.84 =$ 16. $4.16 =$

17. $5.65 =$ 18. $8.375 =$

19. $7.62 =$ 20. $9.44 =$

Change decimals to fractions and write the answers in simplest form.

1. $8.12 =$

2. $2.07 =$

3. $9.37 =$

4. $4.2 =$

5. $6.56 =$

6. $5.06 =$

7. $0.72 =$

8. $3.7 =$

9. $7.22 =$

10. $0.84 =$

11. $2.855 =$

12. $8.11 =$

13. $8.125 =$

14. $1.9 =$

15. $3.62 =$

16. $6.74 =$

17. $5.64 =$

18. $9.875 =$

19. $2.75 =$

20. $6.11 =$

1. $\begin{array}{r} 0.2\ 2 \\ +\ 0.0\ 5 \\ \hline \end{array}$

2. $\begin{array}{r} 0.4\ 6 \\ +\ 0.4\ 8 \\ \hline \end{array}$

3. $\begin{array}{r} 0.5\ 3 \\ +\ 0.3\ 1 \\ \hline \end{array}$

4. $\begin{array}{r} 0.6\ 8 \\ +\ 0.7\ 2 \\ \hline \end{array}$

5. $\begin{array}{r} 0.4\ 2 \\ +\ 0.2\ 1 \\ \hline \end{array}$

6. $\begin{array}{r} 0.5\ 4 \\ +\ 0.1\ 1 \\ \hline \end{array}$

7. $\begin{array}{r} 0.4\ 8 \\ +\ 0.3\ 6 \\ \hline \end{array}$

8. $\begin{array}{r} 0.7\ 3 \\ +\ 0.4\ 1 \\ \hline \end{array}$

9. $\begin{array}{r} 0.7\ 9 \\ +\ 0.6\ 8 \\ \hline \end{array}$

10. $\begin{array}{r} 0.6\ 5 \\ +\ 0.2\ 3 \\ \hline \end{array}$

11. $\begin{array}{r} 0.6\ 6 \\ +\ 0.0\ 6 \\ \hline \end{array}$

12. $\begin{array}{r} 0.6\ 2 \\ +\ 0.3\ 8 \\ \hline \end{array}$

13. $\begin{array}{r} 0.4\ 7 \\ +\ 0.0\ 4 \\ \hline \end{array}$

14. $\begin{array}{r} 0.3\ 9 \\ +\ 0.6\ 2 \\ \hline \end{array}$

15. $\begin{array}{r} 0.8 \\ +\ 0.3\ 4 \\ \hline \end{array}$

16. $\begin{array}{r} 0.0\ 7 \\ +\ 0.4\ 9 \\ \hline \end{array}$

17. $\begin{array}{r} 0.6\ 6 \\ +\ 0.8\ 3 \\ \hline \end{array}$

18. $\begin{array}{r} 0.5\ 9 \\ +\ 0.8\ 6 \\ \hline \end{array}$

19. $\begin{array}{r} 0.2\ 7 \\ +\ 0.0\ 3 \\ \hline \end{array}$

20. $\begin{array}{r} 0.7\ 6 \\ +\ 0.1\ 4 \\ \hline \end{array}$

21. $\begin{array}{r} 0.9\ 7 \\ +\ 0.7\ 6 \\ \hline \end{array}$

22. $\begin{array}{r} 0.4\ 9 \\ +\ 0.7\ 6 \\ \hline \end{array}$

23. $\begin{array}{r} 0.4\ 6 \\ +\ 0.5\ 6 \\ \hline \end{array}$

24. $\begin{array}{r} 0.9\ 1 \\ +\ 0.4\ 3 \\ \hline \end{array}$

1. $+\begin{array}{r} 0.1\ 3 \\ 0.7\ 7 \end{array}$

2. $+\begin{array}{r} 0.5\ 8 \\ 0.3\ 3 \end{array}$

3. $+\begin{array}{r} 0.6\ 3 \\ 0.9 \end{array}$

4. $+\begin{array}{r} 0.7 \\ 0.9\ 8 \end{array}$

5. $+\begin{array}{r} 0.8\ 9 \\ 0.7\ 1 \end{array}$

6. $+\begin{array}{r} 0.1\ 4 \\ 0.2\ 2 \end{array}$

7. $+\begin{array}{r} 0.4\ 1 \\ 0.9\ 7 \end{array}$

8. $+\begin{array}{r} 0.8\ 3 \\ 0.8\ 7 \end{array}$

9. $+\begin{array}{r} 0.1\ 6 \\ 0.8\ 2 \end{array}$

10. $+\begin{array}{r} 0.5\ 7 \\ 0.9\ 3 \end{array}$

11. $+\begin{array}{r} 0.3\ 8 \\ 0.0\ 2 \end{array}$

12. $+\begin{array}{r} 0.3\ 3 \\ 0.7\ 8 \end{array}$

13. $+\begin{array}{r} 0.8\ 6 \\ 0.9\ 3 \end{array}$

14. $+\begin{array}{r} 0.4\ 1 \\ 0.8\ 6 \end{array}$

15. $+\begin{array}{r} 0.9\ 9 \\ 0.7 \end{array}$

16. $+\begin{array}{r} 0.2\ 7 \\ 0.2\ 3 \end{array}$

17. $+\begin{array}{r} 0.8 \\ 0.3\ 7 \end{array}$

18. $+\begin{array}{r} 0.3 \\ 0.1\ 2 \end{array}$

19. $+\begin{array}{r} 0.2\ 9 \\ 0.7\ 9 \end{array}$

20. $+\begin{array}{r} 0.1\ 1 \\ 0.0\ 6 \end{array}$

21. $+\begin{array}{r} 0.6\ 6 \\ 0.6\ 6 \end{array}$

22. $+\begin{array}{r} 0.8\ 8 \\ 0.2 \end{array}$

23. $+\begin{array}{r} 0.9 \\ 0.9\ 1 \end{array}$

24. $+\begin{array}{r} 0.7\ 4 \\ 0.1\ 4 \end{array}$

1.
$$+\begin{array}{r} 2.3\ 7 \\ 1.2\ 2 \end{array}$$

2.
$$+\begin{array}{r} 1.4\ 5 \\ 0.9\ 1 \end{array}$$

3.
$$+\begin{array}{r} 2.2\ 1 \\ 1.0\ 5 \end{array}$$

4.
$$+\begin{array}{r} 0.5 \\ 2.3\ 1 \end{array}$$

5.
$$+\begin{array}{r} 1.7\ 6 \\ 0.5\ 1 \end{array}$$

6.
$$+\begin{array}{r} 2.4\ 9 \\ 1.3\ 9 \end{array}$$

7.
$$+\begin{array}{r} 0.3\ 9 \\ 2.8\ 8 \end{array}$$

8.
$$+\begin{array}{r} 1.3\ 7 \\ 1.1 \end{array}$$

9.
$$+\begin{array}{r} 2.6\ 8 \\ 1.7\ 8 \end{array}$$

10.
$$+\begin{array}{r} 2.8\ 9 \\ 2.6\ 9 \end{array}$$

11.
$$+\begin{array}{r} 1.7\ 7 \\ 1.0\ 2 \end{array}$$

12.
$$+\begin{array}{r} 1.0\ 5 \\ 0.9\ 7 \end{array}$$

13.
$$+\begin{array}{r} 0.1\ 7 \\ 2.2\ 7 \end{array}$$

14.
$$+\begin{array}{r} 0.2\ 3 \\ 2.9\ 8 \end{array}$$

15.
$$+\begin{array}{r} 2.3\ 3 \\ 0.3\ 7 \end{array}$$

16.
$$+\begin{array}{r} 2.0\ 8 \\ 2.1\ 5 \end{array}$$

17.
$$+\begin{array}{r} 2.8 \\ 2.0\ 4 \end{array}$$

18.
$$+\begin{array}{r} 1.7\ 1 \\ 1.4\ 1 \end{array}$$

19.
$$+\begin{array}{r} 2.4\ 4 \\ 2.8\ 4 \end{array}$$

20.
$$+\begin{array}{r} 1.5\ 7 \\ 0.9\ 8 \end{array}$$

21.
$$+\begin{array}{r} 2.9\ 4 \\ 0.7\ 7 \end{array}$$

22.
$$+\begin{array}{r} 0.5\ 1 \\ 2.8\ 4 \end{array}$$

23.
$$+\begin{array}{r} 0.4 \\ 1.1\ 5 \end{array}$$

24.
$$+\begin{array}{r} 1.2\ 3 \\ 0.5\ 4 \end{array}$$

1. $\begin{array}{r} 2.3 \\ + 1.8\ 1 \\ \hline \end{array}$

2. $\begin{array}{r} 2.6\ 5 \\ + 2.8 \\ \hline \end{array}$

3. $\begin{array}{r} 1.9\ 5 \\ + 3.4\ 4 \\ \hline \end{array}$

4. $\begin{array}{r} 3.2\ 9 \\ + 0.8\ 2 \\ \hline \end{array}$

5. $\begin{array}{r} 2.1\ 2 \\ + 0.3\ 9 \\ \hline \end{array}$

6. $\begin{array}{r} 2.2\ 5 \\ + 0.9\ 8 \\ \hline \end{array}$

7. $\begin{array}{r} 2.1 \\ + 1.4\ 3 \\ \hline \end{array}$

8. $\begin{array}{r} 2.1\ 1 \\ + 1.7\ 4 \\ \hline \end{array}$

9. $\begin{array}{r} 0.7\ 4 \\ + 0.1\ 6 \\ \hline \end{array}$

10. $\begin{array}{r} 0.5\ 3 \\ + 2.1\ 8 \\ \hline \end{array}$

11. $\begin{array}{r} 0.1\ 7 \\ + 3.0\ 9 \\ \hline \end{array}$

12. $\begin{array}{r} 3.7\ 1 \\ + 2.6\ 1 \\ \hline \end{array}$

13. $\begin{array}{r} 2.0\ 2 \\ + 3.2\ 5 \\ \hline \end{array}$

14. $\begin{array}{r} 1.4\ 1 \\ + 2.1\ 2 \\ \hline \end{array}$

15. $\begin{array}{r} 3.0\ 4 \\ + 3.9 \\ \hline \end{array}$

16. $\begin{array}{r} 1.1\ 4 \\ + 1.7\ 1 \\ \hline \end{array}$

17. $\begin{array}{r} 1.6\ 2 \\ + 2.9\ 9 \\ \hline \end{array}$

18. $\begin{array}{r} 3.8\ 7 \\ + 3.9 \\ \hline \end{array}$

19. $\begin{array}{r} 2.4\ 8 \\ + 3.1\ 6 \\ \hline \end{array}$

20. $\begin{array}{r} 0.8\ 7 \\ + 2.8\ 1 \\ \hline \end{array}$

21. $\begin{array}{r} 3.5\ 7 \\ + 3.8\ 9 \\ \hline \end{array}$

22. $\begin{array}{r} 2.7\ 1 \\ + 1.0\ 6 \\ \hline \end{array}$

23. $\begin{array}{r} 3.3\ 8 \\ + 2.6 \\ \hline \end{array}$

24. $\begin{array}{r} 2.9\ 7 \\ + 1.4\ 1 \\ \hline \end{array}$

1. $\begin{aligned}&1.5\ 8\\+&\ 4.7\ 1\\\hline\end{aligned}$

2. $\begin{aligned}&1.1\ 1\\+&\ 1.3\ 2\\\hline\end{aligned}$

3. $\begin{aligned}&2.5\ 6\\+&\ 1.9\ 2\\\hline\end{aligned}$

4. $\begin{aligned}&4.4\ 7\\+&\ 2.7\ 4\\\hline\end{aligned}$

5. $\begin{aligned}&0.7\ 3\\+&\ 1.1\ 9\\\hline\end{aligned}$

6. $\begin{aligned}&0.9\ 1\\+&\ 1.7\ 7\\\hline\end{aligned}$

7. $\begin{aligned}&2.5\ 9\\+&\ 1.7\ 7\\\hline\end{aligned}$

8. $\begin{aligned}&2.3\ 2\\+&\ 2.2\ 6\\\hline\end{aligned}$

9. $\begin{aligned}&4.2\ 1\\+&\ 4.6\ 1\\\hline\end{aligned}$

10. $\begin{aligned}&4.9\ 8\\+&\ 1.1\ 9\\\hline\end{aligned}$

11. $\begin{aligned}&3.2\ 6\\+&\ 4.0\ 5\\\hline\end{aligned}$

12. $\begin{aligned}&1.6\ 4\\+&\ 1.8\ 3\\\hline\end{aligned}$

13. $\begin{aligned}&3.0\ 6\\+&\ 2.6\ 9\\\hline\end{aligned}$

14. $\begin{aligned}&3.4\ 7\\+&\ 2.2\ 1\\\hline\end{aligned}$

15. $\begin{aligned}&3.0\ 7\\+&\ 2.9\ 6\\\hline\end{aligned}$

16. $\begin{aligned}&1.6\ 6\\+&\ 0.8\ 7\\\hline\end{aligned}$

17. $\begin{aligned}&2.0\ 5\\+&\ 0.7\ 5\\\hline\end{aligned}$

18. $\begin{aligned}&0.2\ 9\\+&\ 2.9\ 1\\\hline\end{aligned}$

19. $\begin{aligned}&3.5\ 4\\+&\ 3.9\\\hline\end{aligned}$

20. $\begin{aligned}&3.4\ 4\\+&\ 1.3\ 1\\\hline\end{aligned}$

21. $\begin{aligned}&2.4\ 7\\+&\ 3.8\ 9\\\hline\end{aligned}$

22. $\begin{aligned}&2.7\ 8\\+&\ 4.1\ 4\\\hline\end{aligned}$

23. $\begin{aligned}&0.2\\+&\ 3.1\ 5\\\hline\end{aligned}$

24. $\begin{aligned}&2.7\\+&\ 3.9\\\hline\end{aligned}$

1. $+\ \begin{array}{r} 2.2\ 3 \\ 2.7\ 6 \end{array}$

2. $+\ \begin{array}{r} 1.4\ 2 \\ 4.2\ 5 \end{array}$

3. $+\ \begin{array}{r} 4.7 \\ 1.4\ 7 \end{array}$

4. $+\ \begin{array}{r} 1.5\ 1 \\ 4.7\ 5 \end{array}$

5. $+\ \begin{array}{r} 5.6\ 5 \\ 4.0\ 3 \end{array}$

6. $+\ \begin{array}{r} 4.2\ 4 \\ 5.2\ 1 \end{array}$

7. $+\ \begin{array}{r} 3.4\ 9 \\ 3.8\ 8 \end{array}$

8. $+\ \begin{array}{r} 0.9\ 9 \\ 1.2\ 9 \end{array}$

9. $+\ \begin{array}{r} 5.7\ 6 \\ 0.9\ 5 \end{array}$

10. $+\ \begin{array}{r} 3.8\ 2 \\ 2.2\ 4 \end{array}$

11. $+\ \begin{array}{r} 1.5\ 3 \\ 2.8\ 2 \end{array}$

12. $+\ \begin{array}{r} 5.1\ 6 \\ 3.2\ 2 \end{array}$

13. $+\ \begin{array}{r} 0.5 \\ 2.0\ 2 \end{array}$

14. $+\ \begin{array}{r} 3.9\ 6 \\ 5.7\ 8 \end{array}$

15. $+\ \begin{array}{r} 1.0\ 4 \\ 1.6\ 9 \end{array}$

16. $+\ \begin{array}{r} 3.1\ 4 \\ 0.6\ 7 \end{array}$

17. $+\ \begin{array}{r} 3.6\ 4 \\ 5.8\ 7 \end{array}$

18. $+\ \begin{array}{r} 5.2\ 1 \\ 2.1\ 4 \end{array}$

19. $+\ \begin{array}{r} 1.4\ 4 \\ 4.7\ 5 \end{array}$

20. $+\ \begin{array}{r} 4.2 \\ 1.3\ 5 \end{array}$

21. $+\ \begin{array}{r} 5.9\ 8 \\ 2.4 \end{array}$

22. $+\ \begin{array}{r} 4.5 \\ 3.9\ 2 \end{array}$

23. $+\ \begin{array}{r} 4.2\ 2 \\ 4.1\ 9 \end{array}$

24. $+\ \begin{array}{r} 4.9\ 9 \\ 2.9\ 1 \end{array}$

1. $+\begin{array}{r} 2.2\ 1 \\ 3.9 \end{array}$
2. $+\begin{array}{r} 2.5\ 3 \\ 1.3\ 3 \end{array}$
3. $+\begin{array}{r} 5.2\ 2 \\ 3.6\ 9 \end{array}$
4. $+\begin{array}{r} 3.4\ 7 \\ 0.6\ 1 \end{array}$

5. $+\begin{array}{r} 5.5 \\ 2.5\ 1 \end{array}$
6. $+\begin{array}{r} 3.3\ 4 \\ 1.5\ 1 \end{array}$
7. $+\begin{array}{r} 0.8\ 1 \\ 1.3\ 2 \end{array}$
8. $+\begin{array}{r} 1.8\ 8 \\ 5.8\ 7 \end{array}$

9. $+\begin{array}{r} 2.4\ 1 \\ 0.8 \end{array}$
10. $+\begin{array}{r} 1.7\ 7 \\ 2.0\ 8 \end{array}$
11. $+\begin{array}{r} 5.7\ 8 \\ 1.8\ 8 \end{array}$
12. $+\begin{array}{r} 5.3\ 4 \\ 2.4\ 4 \end{array}$

13. $+\begin{array}{r} 1.9 \\ 3.9\ 4 \end{array}$
14. $+\begin{array}{r} 1.1\ 5 \\ 5.9\ 8 \end{array}$
15. $+\begin{array}{r} 1.2\ 8 \\ 4.6\ 6 \end{array}$
16. $+\begin{array}{r} 3.0\ 8 \\ 5.0\ 8 \end{array}$

17. $+\begin{array}{r} 0.8\ 8 \\ 2.5\ 7 \end{array}$
18. $+\begin{array}{r} 4.0\ 4 \\ 2.4\ 8 \end{array}$
19. $+\begin{array}{r} 3.0\ 4 \\ 4.4 \end{array}$
20. $+\begin{array}{r} 4.1\ 4 \\ 3.2\ 6 \end{array}$

21. $+\begin{array}{r} 1.8\ 9 \\ 2.3\ 9 \end{array}$
22. $+\begin{array}{r} 2.8\ 3 \\ 5.8\ 8 \end{array}$
23. $+\begin{array}{r} 3.9\ 7 \\ 1.2\ 4 \end{array}$
24. $+\begin{array}{r} 4.6\ 8 \\ 4.5\ 4 \end{array}$

1. $\begin{array}{r} 0.9\ 1 \\ +\ 1.3 \\ \hline \end{array}$

2. $\begin{array}{r} 4.1\ 2 \\ +\ 3.4\ 9 \\ \hline \end{array}$

3. $\begin{array}{r} 3.7 \\ +\ 1.8\ 6 \\ \hline \end{array}$

4. $\begin{array}{r} 0.5\ 7 \\ +\ 4.6\ 2 \\ \hline \end{array}$

5. $\begin{array}{r} 1.9\ 4 \\ +\ 4.9\ 7 \\ \hline \end{array}$

6. $\begin{array}{r} 3.8\ 3 \\ +\ 2.4\ 3 \\ \hline \end{array}$

7. $\begin{array}{r} 4.9 \\ +\ 2.7\ 8 \\ \hline \end{array}$

8. $\begin{array}{r} 2.1\ 8 \\ +\ 5.8\ 6 \\ \hline \end{array}$

9. $\begin{array}{r} 1.8\ 7 \\ +\ 2.3\ 3 \\ \hline \end{array}$

10. $\begin{array}{r} 2.9\ 8 \\ +\ 4.0\ 7 \\ \hline \end{array}$

11. $\begin{array}{r} 3.7 \\ +\ 1.5\ 1 \\ \hline \end{array}$

12. $\begin{array}{r} 2.2\ 6 \\ +\ 4.9\ 2 \\ \hline \end{array}$

13. $\begin{array}{r} 5.1\ 7 \\ +\ 3.4\ 3 \\ \hline \end{array}$

14. $\begin{array}{r} 2.9\ 5 \\ +\ 4.2 \\ \hline \end{array}$

15. $\begin{array}{r} 1.3\ 5 \\ +\ 3.0\ 7 \\ \hline \end{array}$

16. $\begin{array}{r} 3.7\ 6 \\ +\ 0.6\ 4 \\ \hline \end{array}$

17. $\begin{array}{r} 4.4\ 4 \\ +\ 1.3\ 1 \\ \hline \end{array}$

18. $\begin{array}{r} 3.4\ 6 \\ +\ 1.5\ 5 \\ \hline \end{array}$

19. $\begin{array}{r} 3.6\ 5 \\ +\ 3.3\ 7 \\ \hline \end{array}$

20. $\begin{array}{r} 1.7\ 6 \\ +\ 4.4\ 2 \\ \hline \end{array}$

21. $\begin{array}{r} 4.7 \\ +\ 5.0\ 1 \\ \hline \end{array}$

22. $\begin{array}{r} 2.2\ 3 \\ +\ 4.0\ 2 \\ \hline \end{array}$

23. $\begin{array}{r} 4.3 \\ +\ 1.4\ 2 \\ \hline \end{array}$

24. $\begin{array}{r} 1.9\ 2 \\ +\ 3.5\ 4 \\ \hline \end{array}$

1. $\begin{array}{r} 6.2\ 4 \\ +\ 2.5\ 6 \\ \hline \end{array}$

2. $\begin{array}{r} 5.1\ 4 \\ +\ 2.7\ 2 \\ \hline \end{array}$

3. $\begin{array}{r} 3.5\ 5 \\ +\ 5.3\ 4 \\ \hline \end{array}$

4. $\begin{array}{r} 5.7\ 5 \\ +\ \ 5.7\ 6 \\ \hline \end{array}$

5. $\begin{array}{r} 2.1\ 3 \\ +\ 6.7\ 9 \\ \hline \end{array}$

6. $\begin{array}{r} 6.8\ 5 \\ +\ 1.8\ 6 \\ \hline \end{array}$

7. $\begin{array}{r} 6.2\ 5 \\ +\ 2.5 \\ \hline \end{array}$

8. $\begin{array}{r} 5.8\ 3 \\ +\ \ 6.1\ 8 \\ \hline \end{array}$

9. $\begin{array}{r} 5.4\ 1 \\ +\ 2.4 \\ \hline \end{array}$

10. $\begin{array}{r} 1.1\ 9 \\ +\ 4.5\ 9 \\ \hline \end{array}$

11. $\begin{array}{r} 3.8\ 7 \\ +\ 3.8\ 4 \\ \hline \end{array}$

12. $\begin{array}{r} 1.6\ 7 \\ +\ 5.1 \\ \hline \end{array}$

13. $\begin{array}{r} 6.0\ 8 \\ +\ 1.2\ 6 \\ \hline \end{array}$

14. $\begin{array}{r} 5.5\ 6 \\ +\ \ 5.2\ 5 \\ \hline \end{array}$

15. $\begin{array}{r} 0.9\ 3 \\ +\ 3.6 \\ \hline \end{array}$

16. $\begin{array}{r} 1.4 \\ +\ 2.4\ 7 \\ \hline \end{array}$

17. $\begin{array}{r} 1.5\ 4 \\ +\ 3.2 \\ \hline \end{array}$

18. $\begin{array}{r} 1.4\ 5 \\ +\ 3.3\ 8 \\ \hline \end{array}$

19. $\begin{array}{r} 2.4\ 6 \\ +\ 1.3\ 5 \\ \hline \end{array}$

20. $\begin{array}{r} 1.4\ 4 \\ +\ 5.0\ 4 \\ \hline \end{array}$

21. $\begin{array}{r} 1.6\ 7 \\ +\ 5.9\ 4 \\ \hline \end{array}$

22. $\begin{array}{r} 3.5\ 5 \\ +\ \ 6.5\ 2 \\ \hline \end{array}$

23. $\begin{array}{r} 6.5\ 3 \\ +\ 1.3\ 6 \\ \hline \end{array}$

24. $\begin{array}{r} 2.3\ 5 \\ +\ 1.2\ 1 \\ \hline \end{array}$

1. $\begin{array}{r} 4.6\ 4 \\ +\ 5.1 \\ \hline \end{array}$ 2. $\begin{array}{r} 3.8\ 5 \\ +\ 5.2\ 6 \\ \hline \end{array}$ 3. $\begin{array}{r} 4.8\ 8 \\ +\ 7.5\ 1 \\ \hline \end{array}$ 4. $\begin{array}{r} 2.6\ 8 \\ +\ 4.0\ 9 \\ \hline \end{array}$

5. $\begin{array}{r} 1.0\ 1 \\ +\ 2.6\ 4 \\ \hline \end{array}$ 6. $\begin{array}{r} 1.5 \\ +\ 2.2\ 8 \\ \hline \end{array}$ 7. $\begin{array}{r} 2.8 \\ +\ 1.5\ 3 \\ \hline \end{array}$ 8. $\begin{array}{r} 3.2 \\ +\ 1.1\ 9 \\ \hline \end{array}$

9. $\begin{array}{r} 6.5\ 6 \\ +\ 7.2\ 6 \\ \hline \end{array}$ 10. $\begin{array}{r} 7.4\ 6 \\ +\ 4.7\ 5 \\ \hline \end{array}$ 11. $\begin{array}{r} 4.2 \\ +\ 2.6\ 8 \\ \hline \end{array}$ 12. $\begin{array}{r} 7.1\ 2 \\ +\ 1.3\ 6 \\ \hline \end{array}$

13. $\begin{array}{r} 5.2\ 8 \\ +\ 5.1\ 5 \\ \hline \end{array}$ 14. $\begin{array}{r} 7.1\ 2 \\ +\ 3.6 \\ \hline \end{array}$ 15. $\begin{array}{r} 6.1\ 6 \\ +\ 2.9\ 2 \\ \hline \end{array}$ 16. $\begin{array}{r} 1.4\ 8 \\ +\ 4.4\ 3 \\ \hline \end{array}$

17. $\begin{array}{r} 4.6\ 5 \\ +\ 2.3\ 7 \\ \hline \end{array}$ 18. $\begin{array}{r} 3.9\ 5 \\ +\ 4.6\ 1 \\ \hline \end{array}$ 19. $\begin{array}{r} 5.4\ 6 \\ +\ 6.3 \\ \hline \end{array}$ 20. $\begin{array}{r} 2.9\ 7 \\ +\ 7.4\ 1 \\ \hline \end{array}$

21. $\begin{array}{r} 4.2\ 4 \\ +\ 4.3\ 7 \\ \hline \end{array}$ 22. $\begin{array}{r} 3.9\ 8 \\ +\ 2.8\ 4 \\ \hline \end{array}$ 23. $\begin{array}{r} 7.4\ 3 \\ +\ 6.7\ 2 \\ \hline \end{array}$ 24. $\begin{array}{r} 7.1 \\ +\ 5.8\ 7 \\ \hline \end{array}$

1. $\begin{array}{r} 2.7\ 0 \\ -\ 2.0\ 9 \\ \hline \end{array}$ 2. $\begin{array}{r} 1.8\ 6 \\ -\ 0.4\ 6 \\ \hline \end{array}$ 3. $\begin{array}{r} 2.9\ 7 \\ -\ 0.4\ 4 \\ \hline \end{array}$ 4. $\begin{array}{r} 2.9\ 1 \\ -\ 2.3\ 1 \\ \hline \end{array}$

5. $\begin{array}{r} 2.6\ 5 \\ -\ 0.2\ \\ \hline \end{array}$ 6. $\begin{array}{r} 1.2\ 4 \\ -\ 0.4\ 7 \\ \hline \end{array}$ 7. $\begin{array}{r} 0.4\ 9 \\ -\ 0.3\ 3 \\ \hline \end{array}$ 8. $\begin{array}{r} 1.6\ 3 \\ -\ 0.9\ \\ \hline \end{array}$

9. $\begin{array}{r} 2.0\ 5 \\ -\ 1.9\ 2 \\ \hline \end{array}$ 10. $\begin{array}{r} 2.9\ 7 \\ -\ 0.3\ 3 \\ \hline \end{array}$ 11. $\begin{array}{r} 1.0\ 6 \\ -\ 0.8\ 2 \\ \hline \end{array}$ 12. $\begin{array}{r} 2.2\ 1 \\ -\ 1.0\ 3 \\ \hline \end{array}$

13. $\begin{array}{r} 2.0\ 6 \\ -\ 0.8\ 6 \\ \hline \end{array}$ 14. $\begin{array}{r} 0.3\ 2 \\ -\ 0.1\ 9 \\ \hline \end{array}$ 15. $\begin{array}{r} 1.2\ 6 \\ -\ 0.7\ 3 \\ \hline \end{array}$ 16. $\begin{array}{r} 1.0\ 8 \\ -\ 0.3\ 7 \\ \hline \end{array}$

17. $\begin{array}{r} 2.5\ 1 \\ -\ 0.9\ 8 \\ \hline \end{array}$ 18. $\begin{array}{r} 1.8\ 7 \\ -\ 0.4\ 2 \\ \hline \end{array}$ 19. $\begin{array}{r} 2.5\ 9 \\ -\ 1.2\ \\ \hline \end{array}$ 20. $\begin{array}{r} 1.4\ 7 \\ -\ 0.5\ 1 \\ \hline \end{array}$

21. $\begin{array}{r} 1.8\ 5 \\ -\ 1.1\ \\ \hline \end{array}$ 22. $\begin{array}{r} 2.8\ 0 \\ -\ 1.1\ 6 \\ \hline \end{array}$ 23. $\begin{array}{r} 2.7\ 3 \\ -\ 0.5\ 5 \\ \hline \end{array}$ 24. $\begin{array}{r} 1.9\ 9 \\ -\ 0.9\ 2 \\ \hline \end{array}$

1.
$$\begin{array}{r} 2.8\,5 \\ -\ 0.4\,5 \\ \hline \end{array}$$

2.
$$\begin{array}{r} 2.2\,1 \\ -\ 1.8 \\ \hline \end{array}$$

3.
$$\begin{array}{r} 1.5\,5 \\ -\ 0.3\,9 \\ \hline \end{array}$$

4.
$$\begin{array}{r} 2.2\,1 \\ -\ 2.0\,5 \\ \hline \end{array}$$

5.
$$\begin{array}{r} 2.5\,1 \\ -\ 0.8\,4 \\ \hline \end{array}$$

6.
$$\begin{array}{r} 1.6\,4 \\ -\ 0.0\,9 \\ \hline \end{array}$$

7.
$$\begin{array}{r} 1.3\,2 \\ -\ 0.6\,5 \\ \hline \end{array}$$

8.
$$\begin{array}{r} 2.5\,6 \\ -\ 0.7 \\ \hline \end{array}$$

9.
$$\begin{array}{r} 1.6\,5 \\ -\ 0.4\,4 \\ \hline \end{array}$$

10.
$$\begin{array}{r} 2.9\,6 \\ -\ 2.0\,8 \\ \hline \end{array}$$

11.
$$\begin{array}{r} 2.4\,7 \\ -\ 1.2\,5 \\ \hline \end{array}$$

12.
$$\begin{array}{r} 2.3\,6 \\ -\ 1.9\,9 \\ \hline \end{array}$$

13.
$$\begin{array}{r} 2.9\,2 \\ -\ 0.1\,3 \\ \hline \end{array}$$

14.
$$\begin{array}{r} 2.1\,6 \\ -\ 1.0\,2 \\ \hline \end{array}$$

15.
$$\begin{array}{r} 2.0\,0 \\ -\ 1.4\,1 \\ \hline \end{array}$$

16.
$$\begin{array}{r} 2.0\,1 \\ -\ 0.2\,6 \\ \hline \end{array}$$

17.
$$\begin{array}{r} 2.2\,3 \\ -\ 0.9\,6 \\ \hline \end{array}$$

18.
$$\begin{array}{r} 1.4\,0 \\ -\ 1.1\,7 \\ \hline \end{array}$$

19.
$$\begin{array}{r} 1.2\,8 \\ -\ 0.2\,9 \\ \hline \end{array}$$

20.
$$\begin{array}{r} 1.9\,4 \\ -\ 0.7\,1 \\ \hline \end{array}$$

21.
$$\begin{array}{r} 2.4\,9 \\ -\ 0.8 \\ \hline \end{array}$$

22.
$$\begin{array}{r} 2.0\,3 \\ -\ 0.3\,9 \\ \hline \end{array}$$

23.
$$\begin{array}{r} 1.7\,2 \\ -\ 0.2\,5 \\ \hline \end{array}$$

24.
$$\begin{array}{r} 3.0\,0 \\ -\ 1.3\,7 \\ \hline \end{array}$$

1. $\begin{array}{r} 2.3\,5 \\ -\ 1.6\,2 \\ \hline \end{array}$

2. $\begin{array}{r} 1.2\,3 \\ -\ 0.0\,4 \\ \hline \end{array}$

3. $\begin{array}{r} 2.5\,6 \\ -\ 2.0\,7 \\ \hline \end{array}$

4. $\begin{array}{r} 1.5\,1 \\ -\ 0.0\,8 \\ \hline \end{array}$

5. $\begin{array}{r} 2.3\,7 \\ -\ 0.2\,8 \\ \hline \end{array}$

6. $\begin{array}{r} 1.9\,6 \\ -\ 1.4\,6 \\ \hline \end{array}$

7. $\begin{array}{r} 2.6\,5 \\ -\ 1.6\,1 \\ \hline \end{array}$

8. $\begin{array}{r} 2.4\,1 \\ -\ 1.7\,4 \\ \hline \end{array}$

9. $\begin{array}{r} 1.7\,9 \\ -\ 0.4\,7 \\ \hline \end{array}$

10. $\begin{array}{r} 2.0\,8 \\ -\ 1.2\,7 \\ \hline \end{array}$

11. $\begin{array}{r} 1.3\,4 \\ -\ 1.0\,2 \\ \hline \end{array}$

12. $\begin{array}{r} 2.9\,4 \\ -\ 0.1\,5 \\ \hline \end{array}$

13. $\begin{array}{r} 1.1\,5 \\ -\ 0.5\,2 \\ \hline \end{array}$

14. $\begin{array}{r} 1.7\,3 \\ -\ 1.3\,3 \\ \hline \end{array}$

15. $\begin{array}{r} 2.7\,6 \\ -\ 1.9\,8 \\ \hline \end{array}$

16. $\begin{array}{r} 2.0\,0 \\ -\ 0.4\,2 \\ \hline \end{array}$

17. $\begin{array}{r} 2.4\,1 \\ -\ 1.2\,6 \\ \hline \end{array}$

18. $\begin{array}{r} 1.7\,6 \\ -\ 0.8\,4 \\ \hline \end{array}$

19. $\begin{array}{r} 2.8\,6 \\ -\ 2.6\,5 \\ \hline \end{array}$

20. $\begin{array}{r} 2.6\,0 \\ -\ 1.6\,1 \\ \hline \end{array}$

21. $\begin{array}{r} 2.6\,4 \\ -\ \ \ \ \ 1 \\ \hline \end{array}$

22. $\begin{array}{r} 2.5\,8 \\ -\ 1.3\,2 \\ \hline \end{array}$

23. $\begin{array}{r} 1.7\,6 \\ -\ 0.3\,9 \\ \hline \end{array}$

24. $\begin{array}{r} 2.1\,7 \\ -\ 0.4\,8 \\ \hline \end{array}$

1.   2.9 1
  − 2.2 1
   ‾‾‾‾‾‾

2.   2.0 4
  − 1.6 4
   ‾‾‾‾‾‾

3.   1.2 3
  − 0.8 5
   ‾‾‾‾‾‾

4.   2.6 5
  − 0.4 3
   ‾‾‾‾‾‾

5.   3.8 3
  − 1.8 9
   ‾‾‾‾‾‾

6.   2.3 3
  − 1.5
   ‾‾‾‾‾‾

7.   1.8 9
  − 1.5 2
   ‾‾‾‾‾‾

8.   3.7 9
  − 0.8 7
   ‾‾‾‾‾‾

9.   3.0 5
  − 2.4 6
   ‾‾‾‾‾‾

10.   2.1 6
  − 1.6 5
   ‾‾‾‾‾‾

11.   2.0 3
  − 2.0 2
   ‾‾‾‾‾‾

12.   3.8 4
  − 1.5 3
   ‾‾‾‾‾‾

13.   2.7 0
  − 1.4 1
   ‾‾‾‾‾‾

14.   3.6 8
  − 3.2 8
   ‾‾‾‾‾‾

15.   1.2 0
  − 0.2 7
   ‾‾‾‾‾‾

16.   3.1 2
  − 0.4 8
   ‾‾‾‾‾‾

17.   3.2 5
  − 3
   ‾‾‾‾‾‾

18.   3.1 9
  − 2.0 6
   ‾‾‾‾‾‾

19.   3.0 7
  − 1.8 2
   ‾‾‾‾‾‾

20.   2.4 1
  − 1.1 5
   ‾‾‾‾‾‾

21.   1.9 4
  − 0.5 5
   ‾‾‾‾‾‾

22.   2.5 9
  − 0.8 1
   ‾‾‾‾‾‾

23.   3.6 2
  − 2.4 6
   ‾‾‾‾‾‾

24.   2.0 7
  − 0.7 9
   ‾‾‾‾‾‾

1. $- \begin{array}{r} 3.4\ 1 \\ 2.8\ 4 \\ \hline \end{array}$ 2. $- \begin{array}{r} 3.1\ 7 \\ 2.3\ 7 \\ \hline \end{array}$ 3. $- \begin{array}{r} 3.5\ 0 \\ 0.5\ 8 \\ \hline \end{array}$ 4. $- \begin{array}{r} 2.4\ 2 \\ 0.4\ 8 \\ \hline \end{array}$

5. $- \begin{array}{r} 2.5\ 1 \\ 1.5\ 7 \\ \hline \end{array}$ 6. $- \begin{array}{r} 2.5\ 8 \\ 1.7 \\ \hline \end{array}$ 7. $- \begin{array}{r} 2.8\ 1 \\ 1.8\ 6 \\ \hline \end{array}$ 8. $- \begin{array}{r} 2.8\ 6 \\ 1.7\ 8 \\ \hline \end{array}$

9. $- \begin{array}{r} 3.0\ 4 \\ 1.0\ 8 \\ \hline \end{array}$ 10. $- \begin{array}{r} 3.3\ 4 \\ 0.6\ 1 \\ \hline \end{array}$ 11. $- \begin{array}{r} 1.4\ 7 \\ 0.5\ 7 \\ \hline \end{array}$ 12. $- \begin{array}{r} 2.1\ 5 \\ 0.4\ 2 \\ \hline \end{array}$

13. $- \begin{array}{r} 2.0\ 6 \\ 1.9 \\ \hline \end{array}$ 14. $- \begin{array}{r} 1.8\ 7 \\ 0.1\ 7 \\ \hline \end{array}$ 15. $- \begin{array}{r} 1.4\ 0 \\ 0.7\ 9 \\ \hline \end{array}$ 16. $- \begin{array}{r} 3.4\ 0 \\ 2.4\ 6 \\ \hline \end{array}$

17. $- \begin{array}{r} 2.0\ 4 \\ 0.5\ 8 \\ \hline \end{array}$ 18. $- \begin{array}{r} 2.7\ 0 \\ 2.0\ 5 \\ \hline \end{array}$ 19. $- \begin{array}{r} 3.0\ 0\ 0 \\ 0.1\ 6\ 8 \\ \hline \end{array}$ 20. $- \begin{array}{r} 3.7\ 8 \\ 2.5\ 4 \\ \hline \end{array}$

21. $- \begin{array}{r} 3.3\ 2 \\ 1.6\ 4 \\ \hline \end{array}$ 22. $- \begin{array}{r} 3.1\ 1 \\ 1.9\ 5 \\ \hline \end{array}$ 23. $- \begin{array}{r} 2.6\ 8 \\ 1.7\ 5 \\ \hline \end{array}$ 24. $- \begin{array}{r} 1.8\ 0 \\ 0.5\ 5 \\ \hline \end{array}$

1.
$$\begin{array}{r} 2.9\,0 \\ -\ 1.1\,4 \\ \hline \end{array}$$

2.
$$\begin{array}{r} 3.3\,3 \\ -\ 2.0\,9 \\ \hline \end{array}$$

3.
$$\begin{array}{r} 3.7\,0 \\ -\ 1.9\,6 \\ \hline \end{array}$$

4.
$$\begin{array}{r} 3.5\,1 \\ -\ 1.2\,2 \\ \hline \end{array}$$

5.
$$\begin{array}{r} 2.5\,7 \\ -\ 1.7\,8 \\ \hline \end{array}$$

6.
$$\begin{array}{r} 2.9\,3 \\ -\ 1.1\,6 \\ \hline \end{array}$$

7.
$$\begin{array}{r} 2.0\,6 \\ -\ 0.3\,9 \\ \hline \end{array}$$

8.
$$\begin{array}{r} 3.2\,1 \\ -\ 0.3\,4 \\ \hline \end{array}$$

9.
$$\begin{array}{r} 3.8\,5 \\ -\ 1.5\,2 \\ \hline \end{array}$$

10.
$$\begin{array}{r} 2.0\,7 \\ -\ 1.9\,3 \\ \hline \end{array}$$

11.
$$\begin{array}{r} 4.0\,0 \\ -\ 3.1\,3 \\ \hline \end{array}$$

12.
$$\begin{array}{r} 3.6\,2 \\ -\ 3.1\,6 \\ \hline \end{array}$$

13.
$$\begin{array}{r} 3.6\,0 \\ -\ 1.9\,2 \\ \hline \end{array}$$

14.
$$\begin{array}{r} 1.1\,5 \\ -\ 0.1\,6 \\ \hline \end{array}$$

15.
$$\begin{array}{r} 2.9\,5 \\ -\ 1 \\ \hline \end{array}$$

16.
$$\begin{array}{r} 3.7\,0 \\ -\ 0.3\,8 \\ \hline \end{array}$$

17.
$$\begin{array}{r} 2.8\,5 \\ -\ 0.9 \\ \hline \end{array}$$

18.
$$\begin{array}{r} 1.8\,6 \\ -\ 1.5\,9 \\ \hline \end{array}$$

19.
$$\begin{array}{r} 1.3\,4 \\ -\ 0.6\,6 \\ \hline \end{array}$$

20.
$$\begin{array}{r} 1.7\,8 \\ -\ 1.1\,1 \\ \hline \end{array}$$

21.
$$\begin{array}{r} 3.0\,7 \\ -\ 1.2\,6 \\ \hline \end{array}$$

22.
$$\begin{array}{r} 2.6\,1 \\ -\ 0.7\,5 \\ \hline \end{array}$$

23.
$$\begin{array}{r} 3.5\,3 \\ -\ 2.9\,4 \\ \hline \end{array}$$

24.
$$\begin{array}{r} 3.3\,4 \\ -\ 1.6\,9 \\ \hline \end{array}$$

1.　3.1 2
　　－0.2 7
　　─────

2.　4.8 3
　　－0.8 9
　　─────

3.　3.1 8
　　－0.8 8
　　─────

4.　2.6 1
　　－2.2 8
　　─────

5.　4.8 8
　　－2.5 5
　　─────

6.　3.7 8
　　－2.9 6
　　─────

7.　4.4 6
　　－4.0 7
　　─────

8.　3.8 2
　　－2.6 4
　　─────

9.　1.5 4
　　－0.4 6
　　─────

10.　4.0 0
　　－2.9 8
　　─────

11.　1.7 1
　　－0.9 8
　　─────

12.　4.6 4
　　－0.3 6
　　─────

13.　3.6 6
　　－2.1
　　─────

14.　4.9 6
　　－1.7 6
　　─────

15.　2.2 6
　　－2
　　─────

16.　3.9 1
　　－1.1 2
　　─────

17.　2.5 3
　　－1.1 3
　　─────

18.　2.9 6
　　－2.7 9
　　─────

19.　2.7 5
　　－1.6 1
　　─────

20.　4.9 8
　　－4.1 1
　　─────

21.　4.8 0
　　－2.8 1
　　─────

22.　3.9 4
　　－2.7 6
　　─────

23.　4.2 9
　　－1.6
　　─────

24.　4.2 2
　　－0.2 5
　　─────

1. $\begin{array}{r} 4.7\,2 \\ -\ 3.5\,7 \\ \hline \end{array}$

2. $\begin{array}{r} 2.1\,8 \\ -\ 1.1\,5 \\ \hline \end{array}$

3. $\begin{array}{r} 4.7\,2 \\ -\ 0.1\,4 \\ \hline \end{array}$

4. $\begin{array}{r} 3.6\,0 \\ -\ 1.8\,4 \\ \hline \end{array}$

5. $\begin{array}{r} 4.8\,3 \\ -\ 1.1 \\ \hline \end{array}$

6. $\begin{array}{r} 2.2\,4 \\ -\ 1.9 \\ \hline \end{array}$

7. $\begin{array}{r} 4.0\,6 \\ -\ 1.3\,6 \\ \hline \end{array}$

8. $\begin{array}{r} 4.7\,7 \\ -\ 0.1\,5 \\ \hline \end{array}$

9. $\begin{array}{r} 3.0\,8 \\ -\ 0.6\,3 \\ \hline \end{array}$

10. $\begin{array}{r} 3.7\,9 \\ -\ 2 \\ \hline \end{array}$

11. $\begin{array}{r} 3.4\,1 \\ -\ 0.7\,4 \\ \hline \end{array}$

12. $\begin{array}{r} 4.8\,0 \\ -\ 3.3\,7 \\ \hline \end{array}$

13. $\begin{array}{r} 4.6\,1 \\ -\ 3.1\,7 \\ \hline \end{array}$

14. $\begin{array}{r} 4.5\,5 \\ -\ 1.2\,7 \\ \hline \end{array}$

15. $\begin{array}{r} 4.2\,1 \\ -\ 3.3 \\ \hline \end{array}$

16. $\begin{array}{r} 4.1\,2 \\ -\ 3.4\,3 \\ \hline \end{array}$

17. $\begin{array}{r} 3.2\,9 \\ -\ 2.1\,1 \\ \hline \end{array}$

18. $\begin{array}{r} 3.9\,3 \\ -\ 1.6 \\ \hline \end{array}$

19. $\begin{array}{r} 4.9\,9 \\ -\ 4.0\,1 \\ \hline \end{array}$

20. $\begin{array}{r} 2.9\,3 \\ -\ 2.1\,5 \\ \hline \end{array}$

21. $\begin{array}{r} 3.2\,8 \\ -\ 0.6\,2 \\ \hline \end{array}$

22. $\begin{array}{r} 4.8\,7 \\ -\ 4.2\,6 \\ \hline \end{array}$

23. $\begin{array}{r} 4.4\,2 \\ -\ 3.9\,8 \\ \hline \end{array}$

24. $\begin{array}{r} 3.7\,6 \\ -\ 1.8\,7 \\ \hline \end{array}$

1. $\begin{array}{r} 3.9\,6 \\ -\ 1.9\,1 \\ \hline \end{array}$

2. $\begin{array}{r} 1.8\,0 \\ -\ 0.7\,7 \\ \hline \end{array}$

3. $\begin{array}{r} 5.2\,7 \\ -\ 1.8\,4 \\ \hline \end{array}$

4. $\begin{array}{r} 2.4\,8 \\ -\ 2.1\,1 \\ \hline \end{array}$

5. $\begin{array}{r} 2.5\,7 \\ -\ 2.1\,4 \\ \hline \end{array}$

6. $\begin{array}{r} 1.8\,8 \\ -\ 1.5\,1 \\ \hline \end{array}$

7. $\begin{array}{r} 2.1\,1 \\ -\ 0.4 \\ \hline \end{array}$

8. $\begin{array}{r} 4.2\,3 \\ -\ 0.4\,9 \\ \hline \end{array}$

9. $\begin{array}{r} 5.5\,2 \\ -\ 4.3\,2 \\ \hline \end{array}$

10. $\begin{array}{r} 2.1\,9 \\ -\ 1.2\,3 \\ \hline \end{array}$

11. $\begin{array}{r} 5.9\,0 \\ -\ 3.1\,7 \\ \hline \end{array}$

12. $\begin{array}{r} 2.5\,0 \\ -\ 1.5\,3 \\ \hline \end{array}$

13. $\begin{array}{r} 3.1\,7 \\ -\ 1.0\,9 \\ \hline \end{array}$

14. $\begin{array}{r} 3.2\,7 \\ -\ 1.5\,1 \\ \hline \end{array}$

15. $\begin{array}{r} 5.0\,3 \\ -\ 4.6\,2 \\ \hline \end{array}$

16. $\begin{array}{r} 5.2\,6 \\ -\ 3.3\,1 \\ \hline \end{array}$

17. $\begin{array}{r} 2.2\,8 \\ -\ 1.9\,2 \\ \hline \end{array}$

18. $\begin{array}{r} 4.1\,5 \\ -\ 2.9\,9 \\ \hline \end{array}$

19. $\begin{array}{r} 2.7\,3 \\ -\ 1.0\,3 \\ \hline \end{array}$

20. $\begin{array}{r} 3.8\,4 \\ -\ 2.0\,3 \\ \hline \end{array}$

21. $\begin{array}{r} 2.3\,6 \\ -\ 1.0\,7 \\ \hline \end{array}$

22. $\begin{array}{r} 5.4\,7 \\ -\ 3.6\,2 \\ \hline \end{array}$

23. $\begin{array}{r} 4.8\,9 \\ -\ 3.9\,4 \\ \hline \end{array}$

24. $\begin{array}{r} 5.7\,2 \\ -\ 2.0\,7 \\ \hline \end{array}$

1. $\begin{array}{r} 3.3\,8 \\ -\ 2.0\,9 \\ \hline \end{array}$

2. $\begin{array}{r} 5.6\,4 \\ -\ 2.7\,4 \\ \hline \end{array}$

3. $\begin{array}{r} 2.9\,1 \\ -\ \ \ \ 1 \\ \hline \end{array}$

4. $\begin{array}{r} 5.3\,5 \\ -\ 1.7\,7 \\ \hline \end{array}$

5. $\begin{array}{r} 6.2\,7 \\ -\ 4.4\,4 \\ \hline \end{array}$

6. $\begin{array}{r} 3.1\,6 \\ -\ 2.8\,5 \\ \hline \end{array}$

7. $\begin{array}{r} 3.6\,6 \\ -\ 0.8\,9 \\ \hline \end{array}$

8. $\begin{array}{r} 5.0\,6 \\ -\ 2.2\,2 \\ \hline \end{array}$

9. $\begin{array}{r} 4.1\,6 \\ -\ 0.4\,5 \\ \hline \end{array}$

10. $\begin{array}{r} 4.3\,8 \\ -\ 3.9\,7 \\ \hline \end{array}$

11. $\begin{array}{r} 6.5\,8 \\ -\ 5.9\,7 \\ \hline \end{array}$

12. $\begin{array}{r} 6.9\,7 \\ -\ 4.8\,3 \\ \hline \end{array}$

13. $\begin{array}{r} 6.5\,1 \\ -\ 1.0\,4 \\ \hline \end{array}$

14. $\begin{array}{r} 5.1\,6 \\ -\ 3.8\,9 \\ \hline \end{array}$

15. $\begin{array}{r} 4.2\,4 \\ -\ 3.2 \\ \hline \end{array}$

16. $\begin{array}{r} 1.5\,2 \\ -\ 1.0\,8 \\ \hline \end{array}$

17. $\begin{array}{r} 4.8\,8 \\ -\ 3.3\,2 \\ \hline \end{array}$

18. $\begin{array}{r} 6.2\,4 \\ -\ 5.8\,3 \\ \hline \end{array}$

19. $\begin{array}{r} 5.3\,3 \\ -\ 4.6\,4 \\ \hline \end{array}$

20. $\begin{array}{r} 6.0\,7 \\ -\ 3.9\,2 \\ \hline \end{array}$

21. $\begin{array}{r} 6.9\,5 \\ -\ 2.7\,3 \\ \hline \end{array}$

22. $\begin{array}{r} 3.3\,2 \\ -\ 1.7\,6 \\ \hline \end{array}$

23. $\begin{array}{r} 2.7\,1 \\ -\ 1.2\,7 \\ \hline \end{array}$

24. $\begin{array}{r} 4.9\,9 \\ -\ 0.5\,3 \\ \hline \end{array}$

1. $\begin{array}{r} 0.4\,2 \\ \times\ \ 0.7 \\ \hline \end{array}$
2. $\begin{array}{r} 0.9\,5 \\ \times\ \ 0.8 \\ \hline \end{array}$
3. $\begin{array}{r} 0.5\,4 \\ \times\ \ 0.5 \\ \hline \end{array}$
4. $\begin{array}{r} 0.4\,9 \\ \times\ \ 0.3 \\ \hline \end{array}$

5. $\begin{array}{r} 0.6\,1 \\ \times\ \ 0.9 \\ \hline \end{array}$
6. $\begin{array}{r} 0.8\,2 \\ \times\ \ 0.4 \\ \hline \end{array}$
7. $\begin{array}{r} 0.1\,8 \\ \times\ \ 0.7 \\ \hline \end{array}$
8. $\begin{array}{r} 0.9\,3 \\ \times\ \ 0.5 \\ \hline \end{array}$

9. $\begin{array}{r} 0.7\,4 \\ \times\ \ 0.9 \\ \hline \end{array}$
10. $\begin{array}{r} 0.4\,3 \\ \times\ \ 0.5 \\ \hline \end{array}$
11. $\begin{array}{r} 0.1\,9 \\ \times\ \ 0.9 \\ \hline \end{array}$
12. $\begin{array}{r} 0.4\,4 \\ \times\ \ 0.6 \\ \hline \end{array}$

13. $\begin{array}{r} 0.5\,3 \\ \times\ \ 0.4 \\ \hline \end{array}$
14. $\begin{array}{r} 0.2\,9 \\ \times\ \ 0.3 \\ \hline \end{array}$
15. $\begin{array}{r} 0.4\,9 \\ \times\ \ 0.3 \\ \hline \end{array}$
16. $\begin{array}{r} 0.7\,2 \\ \times\ \ 0.2 \\ \hline \end{array}$

17. $\begin{array}{r} 0.9\,4 \\ \times\ \ 0.2 \\ \hline \end{array}$
18. $\begin{array}{r} 0.8\,4 \\ \times\ \ 0.7 \\ \hline \end{array}$
19. $\begin{array}{r} 0.7\,3 \\ \times\ \ 0.8 \\ \hline \end{array}$
20. $\begin{array}{r} 0.9\,2 \\ \times\ \ 0.6 \\ \hline \end{array}$

1. $\times \begin{array}{r} 2.5 \\ 6.3 \end{array}$

2. $\times \begin{array}{r} 5.2 \\ 1.2 \end{array}$

3. $\times \begin{array}{r} 5.2 \\ 2.7 \end{array}$

4. $\times \begin{array}{r} 7.6 \\ 4.1 \end{array}$

5. $\times \begin{array}{r} 8.5 \\ 3.3 \end{array}$

6. $\times \begin{array}{r} 6.5 \\ 2.6 \end{array}$

7. $\times \begin{array}{r} 5.6 \\ 3.6 \end{array}$

8. $\times \begin{array}{r} 4.3 \\ 9.4 \end{array}$

9. $\times \begin{array}{r} 7.4 \\ 1.4 \end{array}$

10. $\times \begin{array}{r} 3.4 \\ 9.9 \end{array}$

11. $\times \begin{array}{r} 9.6 \\ 4.9 \end{array}$

12. $\times \begin{array}{r} 2.1 \\ 2.1 \end{array}$

1. $\times\ \begin{array}{r} 1.4 \\ 1.7 \end{array}$

2. $\times\ \begin{array}{r} 2.4 \\ 7.2 \end{array}$

3. $\times\ \begin{array}{r} 9.5 \\ 9.3 \end{array}$

4. $\times\ \begin{array}{r} 6.2 \\ 1.8 \end{array}$

5. $\times\ \begin{array}{r} 4.4 \\ 2.9 \end{array}$

6. $\times\ \begin{array}{r} 9.2 \\ 2.5 \end{array}$

7. $\times\ \begin{array}{r} 1.3 \\ 9.8 \end{array}$

8. $\times\ \begin{array}{r} 5.3 \\ 2.5 \end{array}$

9. $\times\ \begin{array}{r} 7.1 \\ 9.4 \end{array}$

10. $\times\ \begin{array}{r} 3.9 \\ 3.5 \end{array}$

11. $\times\ \begin{array}{r} 4.1 \\ 5.2 \end{array}$

12. $\times\ \begin{array}{r} 3.6 \\ 8.8 \end{array}$

1.
$$\begin{array}{r} 1.4\ 1 \\ \times \quad 2.8 \\ \hline \end{array}$$

2.
$$\begin{array}{r} 3.1\ 4 \\ \times \quad 3.7 \\ \hline \end{array}$$

3.
$$\begin{array}{r} 4.2\ 5 \\ \times \quad 1.4 \\ \hline \end{array}$$

4.
$$\begin{array}{r} 2.8\ 2 \\ \times \quad 3.2 \\ \hline \end{array}$$

5.
$$\begin{array}{r} 4.2\ 7 \\ \times \quad 4.6 \\ \hline \end{array}$$

6.
$$\begin{array}{r} 3.5\ 5 \\ \times \quad 3.5 \\ \hline \end{array}$$

7.
$$\begin{array}{r} 3.3\ 3 \\ \times \quad 4.5 \\ \hline \end{array}$$

8.
$$\begin{array}{r} 3.2\ 3 \\ \times \quad 2.1 \\ \hline \end{array}$$

9.
$$\begin{array}{r} 2.2\ 3 \\ \times \quad 2.6 \\ \hline \end{array}$$

10.
$$\begin{array}{r} 0.8\ 6 \\ \times \quad 1.3 \\ \hline \end{array}$$

11.
$$\begin{array}{r} 4.0\ 1 \\ \times \quad 1.3 \\ \hline \end{array}$$

12.
$$\begin{array}{r} 4.3\ 9 \\ \times \quad 1.5 \\ \hline \end{array}$$

Name: Lesson 11-5   Multiplying decimals

1. $\times$ 4.1 2   3.6

2. $\times$ 2.0 8   8.2

3. $\times$ 3.5 7   3.5

4. $\times$ 1.3 9   4.8

5. $\times$ 2.9 2   4.3

6. $\times$ 4.5 1   2.7

7. $\times$ 3.1 6   3.2

8. $\times$ 4.7 8   2.6

9. $\times$ 4.7 5   1.8

10. $\times$ 4.6 5   4.8

11. $\times$ 3.2 4   1.6

12. $\times$ 2.4 1   1.6

120

1.
$$\begin{array}{r} 1.9\,6 \\ \times\ \ 1.4 \\ \hline \end{array}$$

2.
$$\begin{array}{r} 5.2\,1 \\ \times\ \ 2.2 \\ \hline \end{array}$$

3.
$$\begin{array}{r} 5.8\,7 \\ \times\ \ 1.8 \\ \hline \end{array}$$

4.
$$\begin{array}{r} 4.7\,6 \\ \times\ \ 1.9 \\ \hline \end{array}$$

5.
$$\begin{array}{r} 3.4\,5 \\ \times\ \ 8.4 \\ \hline \end{array}$$

6.
$$\begin{array}{r} 2.7\,8 \\ \times\ \ 3.1 \\ \hline \end{array}$$

7.
$$\begin{array}{r} 1.9\,3 \\ \times\ \ 5.5 \\ \hline \end{array}$$

8.
$$\begin{array}{r} 2.8\,2 \\ \times\ \ 5.4 \\ \hline \end{array}$$

9.
$$\begin{array}{r} 3.8\,7 \\ \times\ \ 3.6 \\ \hline \end{array}$$

10.
$$\begin{array}{r} 7.6\,7 \\ \times\ \ 6.1 \\ \hline \end{array}$$

11.
$$\begin{array}{r} 6.4\,3 \\ \times\ \ 2.6 \\ \hline \end{array}$$

12.
$$\begin{array}{r} 4.1\,5 \\ \times\ \ 5.9 \\ \hline \end{array}$$

1. $\times \begin{array}{r} 1.7\ 2 \\ 2.9\ 3 \end{array}$
2. $\times \begin{array}{r} 5.7\ 5 \\ 1.4\ 2 \end{array}$
3. $\times \begin{array}{r} 2.7\ 4 \\ 3.0\ 9 \end{array}$

4. $\times \begin{array}{r} 1.2\ 8 \\ 3.5\ 6 \end{array}$
5. $\times \begin{array}{r} 2.2\ 3 \\ 7.6\ 3 \end{array}$
6. $\times \begin{array}{r} 4.4\ 6 \\ 7.7\ 1 \end{array}$

7. $\times \begin{array}{r} 4.8\ 7 \\ 4.3\ 2 \end{array}$
8. $\times \begin{array}{r} 2.6\ 8 \\ 1.0\ 8 \end{array}$
9. $\times \begin{array}{r} 1.8\ 5 \\ 6.9\ 2 \end{array}$

10. $\times \begin{array}{r} 2.2\ 8 \\ 5.5\ 9 \end{array}$
11. $\times \begin{array}{r} 7.6\ 2 \\ 5.7\ 2 \end{array}$
12. $\times \begin{array}{r} 5.5\ 7 \\ 1.1\ 1 \end{array}$

1. $\times$ 2.1 1
       1.8 5

2. $\times$ 1.1 6
       5.2 4

3. $\times$ 4.7 9
       0.3 6

4. $\times$ 4.4 9
       7.6 2

5. $\times$ 5.3 5
       7.4 2

6. $\times$ 5.9 6
       2.2 5

7. $\times$ 1.3 8
       0.7 4

8. $\times$ 0.8 6
       6.6 5

9. $\times$ 2.4 2
       2.6 6

10. $\times$ 7.0 6
       7.8 5

11. $\times$ 3.7 6
       5.9 2

12. $\times$ 0.8 1
       5.4 2

1.
$$\begin{array}{r} 6.0\ 5 \\ \times\ 3.2\ 3 \\ \hline \end{array}$$

2.
$$\begin{array}{r} 6.4\ 5 \\ \times\ 5.8\ 2 \\ \hline \end{array}$$

3.
$$\begin{array}{r} 2.1\ 2 \\ \times\ 1.8\ 8 \\ \hline \end{array}$$

4.
$$\begin{array}{r} 5.3\ 6 \\ \times\ 5.5\ 1 \\ \hline \end{array}$$

5.
$$\begin{array}{r} 3.5\ 9 \\ \times\ 1.3\ 5 \\ \hline \end{array}$$

6.
$$\begin{array}{r} 1.4\ 5 \\ \times\ 2.0\ 3 \\ \hline \end{array}$$

7.
$$\begin{array}{r} 6.5\ 9 \\ \times\ 7.3\ 1 \\ \hline \end{array}$$

8.
$$\begin{array}{r} 3.1\ 8 \\ \times\ 2.3\ 7 \\ \hline \end{array}$$

9.
$$\begin{array}{r} 3.9\ 7 \\ \times\ 0.6\ 4 \\ \hline \end{array}$$

10.
$$\begin{array}{r} 6.9\ 5 \\ \times\ 5.1\ 6 \\ \hline \end{array}$$

11.
$$\begin{array}{r} 2.7\ 6 \\ \times\ 0.9\ 3 \\ \hline \end{array}$$

12.
$$\begin{array}{r} 5.8\ 4 \\ \times\ 2.2\ 4 \\ \hline \end{array}$$

1.
$$\times \begin{array}{r} 2.5\ 3 \\ 4.9\ 4 \end{array}$$

2.
$$\times \begin{array}{r} 6.5\ 3 \\ 2.5\ 8 \end{array}$$

3.
$$\times \begin{array}{r} 1.4\ 6 \\ 3.8\ 5 \end{array}$$

4.
$$\times \begin{array}{r} 2.3\ 4 \\ 6.7\ 8 \end{array}$$

5.
$$\times \begin{array}{r} 5.1\ 1 \\ 4.5\ 8 \end{array}$$

6.
$$\times \begin{array}{r} 3.5\ 5 \\ 3.8\ 8 \end{array}$$

7.
$$\times \begin{array}{r} 5.5\ 1 \\ 6.0\ 2 \end{array}$$

8.
$$\times \begin{array}{r} 0.9\ 1 \\ 2.9\ 5 \end{array}$$

9.
$$\times \begin{array}{r} 1.4\ 4 \\ 5.5\ 3 \end{array}$$

10.
$$\times \begin{array}{r} 4.1\ 6 \\ 3.6\ 9 \end{array}$$

11.
$$\times \begin{array}{r} 1.1\ 7 \\ 3.0\ 4 \end{array}$$

12.
$$\times \begin{array}{r} 3.0\ 7 \\ 2.9\ 9 \end{array}$$

1. $2\overline{)7.86}$ 2. $2\overline{)9.32}$ 3. $2\overline{)6.32}$ 4. $2\overline{)9.12}$

5. $3\overline{)4.35}$ 6. $3\overline{)9.54}$ 7. $3\overline{)5.55}$ 8. $3\overline{)10.44}$

9. $4\overline{)11.12}$ 10. $4\overline{)11.76}$ 11. $4\overline{)9.36}$ 12. $4\overline{)10.84}$

13. $5\overline{)17.2}$ 14. $5\overline{)18.65}$ 15. $5\overline{)23.1}$ 16. $5\overline{)7.7}$

Name:

Lesson **12-2** Dividing decimals

1. $2 \overline{)9.28}$ 2. $2 \overline{)11.26}$ 3. $2 \overline{)8.58}$ 4. $2 \overline{)7.14}$

5. $3 \overline{)4.08}$ 6. $3 \overline{)11.04}$ 7. $3 \overline{)6.81}$ 8. $3 \overline{)8.04}$

9. $4 \overline{)9.04}$ 10. $4 \overline{)10.64}$ 11. $4 \overline{)6.24}$ 12. $4 \overline{)15.32}$

13. $5 \overline{)14.05}$ 14. $5 \overline{)8.95}$ 15. $5 \overline{)12.35}$ 16. $5 \overline{)23.15}$

127

1. $6\overline{)7.68}$ 2. $6\overline{)16.38}$ 3. $6\overline{)21.18}$ 4. $6\overline{)29.1}$

5. $7\overline{)13.44}$ 6. $7\overline{)15.96}$ 7. $7\overline{)13.93}$ 8. $7\overline{)25.69}$

9. $8\overline{)20.64}$ 10. $8\overline{)19.92}$ 11. $8\overline{)27.76}$ 12. $8\overline{)22.88}$

13. $9\overline{)31.14}$ 14. $9\overline{)28.62}$ 15. $9\overline{)40.77}$ 16. $9\overline{)14.31}$

1. $6 \overline{)21.42}$ 2. $6 \overline{)29.34}$ 3. $6 \overline{)28.08}$ 4. $6 \overline{)10.74}$

5. $7 \overline{)12.39}$ 6. $7 \overline{)26.11}$ 7. $7 \overline{)10.01}$ 8. $7 \overline{)31.99}$

9. $8 \overline{)22.64}$ 10. $8 \overline{)18.88}$ 11. $8 \overline{)26.32}$ 12. $8 \overline{)20.08}$

13. $9 \overline{)35.28}$ 14. $9 \overline{)31.23}$ 15. $9 \overline{)42.21}$ 16. $9 \overline{)17.91}$

1. $2.1 \overline{)7.56}$ 2. $1.7 \overline{)2.38}$ 3. $2.9 \overline{)8.99}$ 4. $3.6 \overline{)6.48}$

5. $1.9 \overline{)6.65}$ 6. $3.4 \overline{)9.52}$ 7. $2.1 \overline{)5.46}$ 8. $1.8 \overline{)7.74}$

9. $1.8 \overline{)2.88}$ 10. $1.4 \overline{)5.32}$ 11. $3.3 \overline{)12.21}$ 12. $2.6 \overline{)10.14}$

13. $3.4 \overline{)9.18}$ 14. $2.9 \overline{)4.93}$ 15. $2.4 \overline{)10.56}$ 16. $4.6 \overline{)6.44}$

1. $1.5\overline{)2.25}$ 2. $1.8\overline{)8.82}$ 3. $2.4\overline{)8.16}$ 4. $2.7\overline{)7.02}$

5. $1.9\overline{)3.04}$ 6. $3.3\overline{)12.87}$ 7. $3.7\overline{)12.58}$ 8. $3.1\overline{)13.95}$

9. $3.7\overline{)15.17}$ 10. $4.4\overline{)14.08}$ 11. $1.9\overline{)9.31}$ 12. $4.3\overline{)18.49}$

13. $2.4\overline{)6.96}$ 14. $1.9\overline{)3.42}$ 15. $4.5\overline{)16.65}$ 16. $4.7\overline{)12.69}$

1. $3.4\overline{)16.32}$ 2. $3.2\overline{)12.48}$ 3. $4.1\overline{)15.17}$ 4. $4.7\overline{)22.56}$

5. $1.9\overline{)10.64}$ 6. $4.5\overline{)15.75}$ 7. $2.9\overline{)11.31}$ 8. $3.4\overline{)14.62}$

9. $4.7\overline{)12.69}$ 10. $4.8\overline{)8.16}$ 11. $4.8\overline{)19.68}$ 12. $3.6\overline{)15.12}$

13. $3.2\overline{)15.36}$ 14. $3.8\overline{)18.62}$ 15. $3.9\overline{)17.55}$ 16. $4.9\overline{)23.03}$

1. $4.5\overline{)23.85}$ 2. $4.1\overline{)21.73}$ 3. $4.1\overline{)22.14}$ 4. $2.9\overline{)12.18}$

5. $3.7\overline{)17.02}$ 6. $4.9\overline{)19.11}$ 7. $3.8\overline{)16.72}$ 8. $4.6\overline{)23.46}$

9. $3.4\overline{)20.06}$ 10. $5.5\overline{)25.85}$ 11. $2.9\overline{)15.37}$ 12. $4.8\overline{)21.12}$

13. $5.7\overline{)19.95}$ 14. $3.7\overline{)20.35}$ 15. $3.1\overline{)14.57}$ 16. $5.3\overline{)29.15}$

1. $0.8\overline{)2.08}$ 2. $0.6\overline{)2.28}$ 3. $0.3\overline{)1.44}$ 4. $0.4\overline{)2.16}$

5. $0.4\overline{)1.52}$ 6. $0.9\overline{)5.04}$ 7. $0.6\overline{)2.64}$ 8. $0.3\overline{)1.17}$

9. $0.2\overline{)0.98}$ 10. $0.8\overline{)3.44}$ 11. $0.7\overline{)4.13}$ 12. $0.8\overline{)3.68}$

13. $0.5\overline{)3.35}$ 14. $0.7\overline{)3.22}$ 15. $0.5\overline{)2.4}$ 16. $0.7\overline{)2.24}$

1. $0.7\overline{)4.76}$ 2. $0.5\overline{)3.95}$ 3. $0.9\overline{)7.11}$ 4. $0.3\overline{)2.88}$

5. $0.4\overline{)3.12}$ 6. $0.3\overline{)2.52}$ 7. $0.6\overline{)2.94}$ 8. $0.7\overline{)4.55}$

9. $0.7\overline{)4.83}$ 10. $0.7\overline{)4.41}$ 11. $0.8\overline{)5.84}$ 12. $0.8\overline{)7.04}$

13. $0.4\overline{)3.04}$ 14. $0.5\overline{)4.2}$ 15. $0.6\overline{)5.58}$ 16. $0.9\overline{)8.91}$

1.
$$\begin{array}{r} 1033 \\ \times\ 25 \\ \hline 5165 \\ 2066 \\ \hline 25825 \end{array}$$

2.
$$\begin{array}{r} 1344 \\ \times\ 14 \\ \hline 5376 \\ 1344 \\ \hline 18816 \end{array}$$

3.
$$\begin{array}{r} 1629 \\ \times\ 53 \\ \hline 4887 \\ 8145 \\ \hline 86337 \end{array}$$

4.
$$\begin{array}{r} 1856 \\ \times\ 16 \\ \hline 11136 \\ 1856 \\ \hline 29696 \end{array}$$

5.
$$\begin{array}{r} 1528 \\ \times\ 46 \\ \hline 9168 \\ 6112 \\ \hline 70288 \end{array}$$

6.
$$\begin{array}{r} 1432 \\ \times\ 18 \\ \hline 11456 \\ 1432 \\ \hline 25776 \end{array}$$

7.
$$\begin{array}{r} 1661 \\ \times\ 35 \\ \hline 8305 \\ 4983 \\ \hline 58135 \end{array}$$

8.
$$\begin{array}{r} 1965 \\ \times\ 32 \\ \hline 3930 \\ 5895 \\ \hline 62880 \end{array}$$

9.
$$\begin{array}{r} 1288 \\ \times\ 49 \\ \hline 11592 \\ 5152 \\ \hline 63112 \end{array}$$

10.
$$\begin{array}{r} 1145 \\ \times\ 19 \\ \hline 10305 \\ 1145 \\ \hline 21755 \end{array}$$

11.
$$\begin{array}{r} 1685 \\ \times\ 12 \\ \hline 3370 \\ 1685 \\ \hline 20220 \end{array}$$

12.
$$\begin{array}{r} 1857 \\ \times\ 13 \\ \hline 5571 \\ 1857 \\ \hline 24141 \end{array}$$

1.
$$\begin{array}{r} 2843 \\ \times\ 28 \\ \hline 22744 \\ 5686 \\ \hline 79604 \end{array}$$

2.
$$\begin{array}{r} 1532 \\ \times\ 45 \\ \hline 7660 \\ 6128 \\ \hline 68940 \end{array}$$

3.
$$\begin{array}{r} 2451 \\ \times\ 21 \\ \hline 2451 \\ 4902 \\ \hline 51471 \end{array}$$

4.
$$\begin{array}{r} 1278 \\ \times\ 37 \\ \hline 8946 \\ 3834 \\ \hline 47286 \end{array}$$

5.
$$\begin{array}{r} 2380 \\ \times\ 19 \\ \hline 21420 \\ 2380 \\ \hline 45220 \end{array}$$

6.
$$\begin{array}{r} 2826 \\ \times\ 33 \\ \hline 8478 \\ 8478 \\ \hline 93258 \end{array}$$

7.
$$\begin{array}{r} 2123 \\ \times\ 25 \\ \hline 10615 \\ 4246 \\ \hline 53075 \end{array}$$

8.
$$\begin{array}{r} 1583 \\ \times\ 55 \\ \hline 7915 \\ 7915 \\ \hline 87065 \end{array}$$

9.
$$\begin{array}{r} 1682 \\ \times\ 58 \\ \hline 13456 \\ 8410 \\ \hline 97556 \end{array}$$

10.
$$\begin{array}{r} 2584 \\ \times\ 19 \\ \hline 23256 \\ 2584 \\ \hline 49096 \end{array}$$

11.
$$\begin{array}{r} 1302 \\ \times\ 11 \\ \hline 1302 \\ 1302 \\ \hline 14322 \end{array}$$

12.
$$\begin{array}{r} 2486 \\ \times\ 28 \\ \hline 19888 \\ 4972 \\ \hline 69608 \end{array}$$

1.
$$\begin{array}{r} 3154 \\ \times\ 24 \\ \hline 12616 \\ 6308 \\ \hline 75696 \end{array}$$

2.
$$\begin{array}{r} 1816 \\ \times\ 32 \\ \hline 3632 \\ 5448 \\ \hline 58112 \end{array}$$

3.
$$\begin{array}{r} 1360 \\ \times\ 42 \\ \hline 2720 \\ 5440 \\ \hline 57120 \end{array}$$

4.
$$\begin{array}{r} 3678 \\ \times\ 29 \\ \hline 33102 \\ 7356 \\ \hline 106662 \end{array}$$

5.
$$\begin{array}{r} 2638 \\ \times\ 39 \\ \hline 23742 \\ 7914 \\ \hline 102882 \end{array}$$

6.
$$\begin{array}{r} 3446 \\ \times\ 15 \\ \hline 17230 \\ 3446 \\ \hline 51690 \end{array}$$

7.
$$\begin{array}{r} 3570 \\ \times\ 19 \\ \hline 32130 \\ 3570 \\ \hline 67830 \end{array}$$

8.
$$\begin{array}{r} 1965 \\ \times\ 45 \\ \hline 9825 \\ 7860 \\ \hline 88425 \end{array}$$

9.
$$\begin{array}{r} 3096 \\ \times\ 18 \\ \hline 24768 \\ 3096 \\ \hline 55728 \end{array}$$

10.
$$\begin{array}{r} 2489 \\ \times\ 17 \\ \hline 17423 \\ 2489 \\ \hline 42313 \end{array}$$

11.
$$\begin{array}{r} 2036 \\ \times\ 26 \\ \hline 12216 \\ 4072 \\ \hline 52936 \end{array}$$

12.
$$\begin{array}{r} 2503 \\ \times\ 29 \\ \hline 22527 \\ 5006 \\ \hline 72587 \end{array}$$

1.
$$\begin{array}{r} 3782 \\ \times\ 27 \\ \hline 26474 \\ 7564 \\ \hline 102114 \end{array}$$

2.
$$\begin{array}{r} 3166 \\ \times\ 16 \\ \hline 18996 \\ 3166 \\ \hline 50656 \end{array}$$

3.
$$\begin{array}{r} 1707 \\ \times\ 17 \\ \hline 11949 \\ 1707 \\ \hline 29019 \end{array}$$

4.
$$\begin{array}{r} 1878 \\ \times\ 39 \\ \hline 16902 \\ 5634 \\ \hline 73242 \end{array}$$

5.
$$\begin{array}{r} 3217 \\ \times\ 24 \\ \hline 12868 \\ 6434 \\ \hline 77208 \end{array}$$

6.
$$\begin{array}{r} 2795 \\ \times\ 26 \\ \hline 16770 \\ 5590 \\ \hline 72670 \end{array}$$

7.
$$\begin{array}{r} 2114 \\ \times\ 42 \\ \hline 4228 \\ 8456 \\ \hline 88788 \end{array}$$

8.
$$\begin{array}{r} 1496 \\ \times\ 67 \\ \hline 10472 \\ 8976 \\ \hline 100232 \end{array}$$

9.
$$\begin{array}{r} 3711 \\ \times\ 14 \\ \hline 14844 \\ 3711 \\ \hline 51954 \end{array}$$

10.
$$\begin{array}{r} 3863 \\ \times\ 16 \\ \hline 23178 \\ 3863 \\ \hline 61808 \end{array}$$

11.
$$\begin{array}{r} 2491 \\ \times\ 34 \\ \hline 9964 \\ 7473 \\ \hline 84694 \end{array}$$

12.
$$\begin{array}{r} 2451 \\ \times\ 16 \\ \hline 14706 \\ 2451 \\ \hline 39216 \end{array}$$

Lesson 1-5 Multiplying two and four digit numbers

1.
$$\begin{array}{r} 1811 \\ \times \quad 22 \\ \hline 3622 \\ 3622 \\ \hline 39842 \end{array}$$

2.
$$\begin{array}{r} 2846 \\ \times \quad 35 \\ \hline 14230 \\ 8538 \\ \hline 99610 \end{array}$$

3.
$$\begin{array}{r} 2794 \\ \times \quad 25 \\ \hline 13970 \\ 5588 \\ \hline 69850 \end{array}$$

4.
$$\begin{array}{r} 1996 \\ \times \quad 31 \\ \hline 1996 \\ 5988 \\ \hline 61876 \end{array}$$

5.
$$\begin{array}{r} 2623 \\ \times \quad 16 \\ \hline 15738 \\ 2623 \\ \hline 41968 \end{array}$$

6.
$$\begin{array}{r} 2374 \\ \times \quad 34 \\ \hline 9496 \\ 7122 \\ \hline 80716 \end{array}$$

7.
$$\begin{array}{r} 3219 \\ \times \quad 27 \\ \hline 22533 \\ 6438 \\ \hline 86913 \end{array}$$

8.
$$\begin{array}{r} 3311 \\ \times \quad 21 \\ \hline 3311 \\ 6622 \\ \hline 69531 \end{array}$$

9.
$$\begin{array}{r} 3338 \\ \times \quad 13 \\ \hline 10014 \\ 3338 \\ \hline 43394 \end{array}$$

10.
$$\begin{array}{r} 2742 \\ \times \quad 29 \\ \hline 24678 \\ 5484 \\ \hline 79518 \end{array}$$

11.
$$\begin{array}{r} 3139 \\ \times \quad 25 \\ \hline 15695 \\ 6278 \\ \hline 78475 \end{array}$$

12.
$$\begin{array}{r} 3649 \\ \times \quad 16 \\ \hline 21894 \\ 3649 \\ \hline 58384 \end{array}$$

Lesson 1-6 Multiplying two and four digit numbers

1.
$$\begin{array}{r} 2870 \\ \times \quad 54 \\ \hline 11480 \\ 14350 \\ \hline 154980 \end{array}$$

2.
$$\begin{array}{r} 3921 \\ \times \quad 35 \\ \hline 19605 \\ 11763 \\ \hline 137235 \end{array}$$

3.
$$\begin{array}{r} 2156 \\ \times \quad 64 \\ \hline 8624 \\ 12936 \\ \hline 137984 \end{array}$$

4.
$$\begin{array}{r} 3450 \\ \times \quad 82 \\ \hline 6900 \\ 27600 \\ \hline 282900 \end{array}$$

5.
$$\begin{array}{r} 4327 \\ \times \quad 61 \\ \hline 4327 \\ 25962 \\ \hline 263947 \end{array}$$

6.
$$\begin{array}{r} 3326 \\ \times \quad 56 \\ \hline 19956 \\ 16630 \\ \hline 186256 \end{array}$$

7.
$$\begin{array}{r} 2789 \\ \times \quad 62 \\ \hline 5578 \\ 16734 \\ \hline 172918 \end{array}$$

8.
$$\begin{array}{r} 3555 \\ \times \quad 42 \\ \hline 7110 \\ 14220 \\ \hline 149310 \end{array}$$

9.
$$\begin{array}{r} 2673 \\ \times \quad 79 \\ \hline 24057 \\ 18711 \\ \hline 211167 \end{array}$$

10.
$$\begin{array}{r} 2754 \\ \times \quad 69 \\ \hline 24786 \\ 16524 \\ \hline 190026 \end{array}$$

11.
$$\begin{array}{r} 4066 \\ \times \quad 65 \\ \hline 20330 \\ 24396 \\ \hline 264290 \end{array}$$

12.
$$\begin{array}{r} 3468 \\ \times \quad 48 \\ \hline 27744 \\ 13872 \\ \hline 166464 \end{array}$$

Lesson 1-7 Multiplying two and four digit numbers

1.
$$\begin{array}{r} 4534 \\ \times \quad 46 \\ \hline 27204 \\ 18136 \\ \hline 208564 \end{array}$$

2.
$$\begin{array}{r} 4904 \\ \times \quad 84 \\ \hline 19616 \\ 39232 \\ \hline 411936 \end{array}$$

3.
$$\begin{array}{r} 4124 \\ \times \quad 49 \\ \hline 37116 \\ 16496 \\ \hline 202076 \end{array}$$

4.
$$\begin{array}{r} 2850 \\ \times \quad 54 \\ \hline 11400 \\ 14250 \\ \hline 153900 \end{array}$$

5.
$$\begin{array}{r} 3088 \\ \times \quad 46 \\ \hline 18528 \\ 12352 \\ \hline 142048 \end{array}$$

6.
$$\begin{array}{r} 2698 \\ \times \quad 82 \\ \hline 5396 \\ 21584 \\ \hline 221236 \end{array}$$

7.
$$\begin{array}{r} 3945 \\ \times \quad 38 \\ \hline 31560 \\ 11835 \\ \hline 149910 \end{array}$$

8.
$$\begin{array}{r} 2125 \\ \times \quad 85 \\ \hline 10625 \\ 17000 \\ \hline 180625 \end{array}$$

9.
$$\begin{array}{r} 3224 \\ \times \quad 43 \\ \hline 9672 \\ 12896 \\ \hline 138632 \end{array}$$

10.
$$\begin{array}{r} 3562 \\ \times \quad 57 \\ \hline 24934 \\ 17810 \\ \hline 203034 \end{array}$$

11.
$$\begin{array}{r} 4419 \\ \times \quad 29 \\ \hline 39771 \\ 8838 \\ \hline 128151 \end{array}$$

12.
$$\begin{array}{r} 3492 \\ \times \quad 68 \\ \hline 27936 \\ 20952 \\ \hline 237456 \end{array}$$

Lesson 1-8 Multiplying two and four digit numbers

1.
$$\begin{array}{r} 2248 \\ \times \quad 63 \\ \hline 6744 \\ 13488 \\ \hline 141624 \end{array}$$

2.
$$\begin{array}{r} 4227 \\ \times \quad 38 \\ \hline 33816 \\ 12681 \\ \hline 160626 \end{array}$$

3.
$$\begin{array}{r} 5743 \\ \times \quad 23 \\ \hline 17229 \\ 11486 \\ \hline 132089 \end{array}$$

4.
$$\begin{array}{r} 5145 \\ \times \quad 27 \\ \hline 36015 \\ 10290 \\ \hline 138915 \end{array}$$

5.
$$\begin{array}{r} 5330 \\ \times \quad 32 \\ \hline 10660 \\ 15990 \\ \hline 170560 \end{array}$$

6.
$$\begin{array}{r} 3825 \\ \times \quad 71 \\ \hline 3825 \\ 26775 \\ \hline 271575 \end{array}$$

7.
$$\begin{array}{r} 3689 \\ \times \quad 38 \\ \hline 29512 \\ 11067 \\ \hline 140182 \end{array}$$

8.
$$\begin{array}{r} 2741 \\ \times \quad 57 \\ \hline 19187 \\ 13705 \\ \hline 156237 \end{array}$$

9.
$$\begin{array}{r} 5556 \\ \times \quad 37 \\ \hline 38892 \\ 16668 \\ \hline 205572 \end{array}$$

10.
$$\begin{array}{r} 3958 \\ \times \quad 71 \\ \hline 3958 \\ 27706 \\ \hline 281018 \end{array}$$

11.
$$\begin{array}{r} 5171 \\ \times \quad 43 \\ \hline 15513 \\ 20684 \\ \hline 222353 \end{array}$$

12.
$$\begin{array}{r} 4616 \\ \times \quad 89 \\ \hline 41544 \\ 36928 \\ \hline 410824 \end{array}$$

1.
```
  × 1 9 8 6
        9 6
  1 1 9 1 6
  1 7 8 7 4
1 9 0 6 5 6
```

2.
```
  × 3 6 6 4
        4 7
  2 5 6 4 8
  1 4 6 5 6
1 7 2 2 0 8
```

3.
```
  × 5 2 7 2
        3 8
  4 2 1 7 6
  1 5 8 1 6
2 0 0 3 3 6
```

4.
```
  × 4 1 6 4
        4 6
  2 4 9 8 4
  1 6 6 5 6
1 9 1 5 4 4
```

5.
```
  × 4 9 9 1
        5 8
  3 9 9 2 8
  2 4 9 5 5
2 8 9 4 7 8
```

6.
```
  × 4 4 3 9
        5 6
  2 6 6 3 4
  2 2 1 9 5
2 4 8 5 8 4
```

7.
```
  × 2 2 4 6
        7 5
  1 1 2 3 0
  1 5 7 2 2
1 6 8 4 5 0
```

8.
```
  × 5 0 1 3
        7 2
  1 0 0 2 6
  3 5 0 9 1
3 6 0 9 3 6
```

9.
```
  × 5 7 5 4
        4 4
  2 3 0 1 6
  2 3 0 1 6
2 5 3 1 7 6
```

10.
```
  × 3 8 3 2
        4 7
  2 6 8 2 4
  1 5 3 2 8
1 8 0 1 0 4
```

11.
```
  × 4 2 8 3
        3 6
  2 5 6 9 8
  1 2 8 4 9
1 5 4 1 8 8
```

12.
```
  × 4 1 9 7
        5 1
  4 1 9 7
  2 0 9 8 5
2 1 4 0 4 7
```

1.
```
  × 3 6 6 1
        6 6
  2 1 9 6 6
  2 1 9 6 6
2 4 1 6 2 6
```

2.
```
  × 2 6 9 7
        9 2
  5 3 9 4
  2 4 2 7 3
2 4 8 1 2 4
```

3.
```
  × 3 2 5 6
        6 1
  3 2 5 6
  1 9 5 3 6
1 9 8 6 1 6
```

4.
```
  × 5 6 3 7
        5 3
  1 6 9 1 1
  2 8 1 8 5
2 9 8 7 6 1
```

5.
```
  × 5 6 0 4
        3 7
  3 9 2 2 8
  1 6 8 1 2
2 0 7 3 4 8
```

6.
```
  × 4 5 8 2
        4 9
  4 1 2 3 8
  1 8 3 2 8
2 2 4 5 1 8
```

7.
```
  × 2 3 9 6
        5 3
  7 1 8 8
  1 1 9 8 0
1 2 6 9 8 8
```

8.
```
  × 6 7 4 7
        6 3
  2 0 2 4 1
  4 0 4 8 2
4 2 5 0 6 1
```

9.
```
  × 4 6 7 6
        5 4
  1 8 7 0 4
  2 3 3 8 0
2 5 2 5 0 4
```

10.
```
  × 5 2 4 2
        4 1
  5 2 4 2
  2 0 9 6 8
2 1 4 9 2 2
```

11.
```
  × 4 7 6 5
        6 9
  4 2 8 8 5
  2 8 5 9 0
3 2 8 7 8 5
```

12.
```
  × 5 8 5 9
        7 8
  4 6 8 7 2
  4 1 0 1 3
4 5 7 0 0 2
```

1.
```
         77
  22 ) 1694
       1540
        154
        154
          0
```

2.
```
        181
  15 ) 2716
       1500
       1216
       1200
         16
         15
          1
```

3.
```
        144
  16 ) 2313
       1600
        713
        640
         73
         64
          9
```

5.
```
        118
  17 ) 2006
       1700
        306
        170
        136
        136
          0
```

6.
```
         91
  29 ) 2642
       2610
         32
         29
          3
```

7.
```
         93
  34 ) 3177
       3060
        117
        102
         15
```

9.
```
        159
  13 ) 2079
       1300
        779
        650
        129
        117
         12
```

10.
```
        206
  18 ) 3724
       3600
        124
        108
         16
```

11.
```
        133
  22 ) 2941
       2200
        741
        660
         81
         66
         15
```

13.
```
         94
  33 ) 3129
       2970
        159
        132
         27
```

14.
```
        111
  19 ) 2115
       1900
        215
        190
         25
         19
          6
```

15.
```
        199
  17 ) 3387
       1700
       1687
       1530
        157
        153
          4
```

1.
```
        118
  11 ) 1306
       1100
        206
        110
         96
         88
          8
```

2.
```
        160
  15 ) 2405
       1500
        905
        900
          5
```

3.
```
        134
  19 ) 2557
       1900
        657
        570
         87
         76
         11
```

6.
```
        133
  27 ) 3599
       2700
        899
        810
         89
         81
          8
```

5.
```
        166
  21 ) 3493
       2100
       1393
       1260
        133
        126
          7
```

7.
```
        107
  31 ) 3333
       3100
        233
        217
         16
```

9.
```
        126
  17 ) 2147
       1700
        447
        340
        107
        102
          5
```

10.
```
        114
  19 ) 2169
       1900
        269
        190
         79
         76
          3
```

11.
```
        133
  26 ) 3478
       2600
        878
        780
         98
         78
         20
```

13.
```
        127
  25 ) 3181
       2500
        681
        500
        181
        175
          6
```

14.
```
        142
  13 ) 1850
       1300
        550
        520
         30
         26
          4
```

15.
```
        138
  17 ) 2352
       1700
        652
        510
        142
        136
          6
```

Lesson 2-3   Dividing by two digit numbers with remainders

1.
```
       114
   16)1831
      1600
       231
       160
        71
        64
         7
```

2.
```
       186
   14)2617
      1400
      1217
      1120
        97
        84
        13
```

3.
```
       168
   12)2019
      1200
       819
       720
        99
        96
         3
```

5.
```
       123
   13)1605
      1300
       305
       260
        45
        39
         6
```

6.
```
       192
   19)3664
      1900
      1764
      1710
        54
        38
        16
```

7.
```
       129
   19)2459
      1900
       559
       380
       179
       171
         8
```

9.
```
       110
   25)2764
      2500
       264
       250
        14
```

10.
```
       114
   22)2515
      2200
       315
       220
        95
        88
         7
```

11.
```
       143
   23)3292
      2300
       992
       920
        72
        69
         3
```

13.
```
       138
   27)3748
      2700
      1048
       810
       238
       216
        22
```

14.
```
       141
   24)3394
      2400
       994
       960
        34
        24
        10
```

15.
```
       170
   17)2901
      1700
      1201
      1190
        11
```

Lesson 2-4   Dividing by two digit numbers with remainders

1.
```
       188
   14)2638
      1400
      1238
      1120
       118
       112
         6
```

2.
```
       145
   29)4212
      2900
      1312
      1160
       152
       145
         7
```

3.
```
       110
   22)2431
      2200
       231
       220
        11
```

7.
```
       187
   17)3188
      1700
      1488
      1360
       128
       119
         9
```

5.
```
       118
   27)3187
      2700
       487
       270
       217
       216
         1
```

6.
```
       162
   17)2770
      1700
      1070
      1020
        50
        34
        16
```

9.
```
       135
   31)4201
      3100
      1101
       930
       171
       155
        16
```

10.
```
       132
   32)4239
      3200
      1039
       960
        79
        64
        15
```

11.
```
       124
   34)4235
      3400
       835
       680
       155
       136
        19
```

13.
```
       208
   23)4800
      4600
       200
       184
        16
```

14.
```
       160
   25)4014
      2500
      1514
      1500
        14
```

15.
```
       161
   21)3400
      2100
      1300
      1260
        40
        21
        19
```

Lesson 2-5   Dividing by two digit numbers with remainders

1.
```
       288
   12)3467
      2400
      1067
       960
       107
        96
        11
```

2.
```
       186
   21)3926
      2100
      1826
      1680
       146
       126
        20
```

3.
```
       172
   26)4491
      2600
      1891
      1820
        71
        52
        19
```

5.
```
       217
   15)3264
      3000
       264
       150
       114
       105
         9
```

6.
```
       135
   18)2430
      1800
       630
       540
        90
        90
         0
```

7.
```
       124
   17)2108
      1700
       408
       340
        68
        68
         0
```

9.
```
       166
   14)2337
      1400
       937
       840
        97
        84
        13
```

10.
```
       183
   19)3490
      1900
      1590
      1520
        70
        57
        13
```

11.
```
       106
   21)2241
      2100
       141
       126
        15
```

13.
```
       123
   17)2094
      1700
       394
       340
        54
        51
         3
```

14.
```
       103
   39)4021
      3900
       121
       117
         4
```

15.
```
       180
   15)2706
      1500
      1206
      1200
         6
```

Lesson 2-6   Dividing by two digit numbers with remainders

1.
```
       161
   25)4043
      2500
      1543
      1500
        43
        25
        18
```

2.
```
       117
   42)4931
      4200
       731
       420
       311
       294
        17
```

3.
```
       112
   36)4060
      3600
       460
       360
       100
        72
        28
```

5.
```
       293
   19)5567
      3800
      1767
      1710
        57
        57
         0
```

6.
```
       305
   12)3661
      3600
        61
        60
         1
```

7.
```
       135
   28)3781
      2800
       981
       840
       141
       140
         1
```

9.
```
       198
   11)2188
      1100
      1088
       990
        98
        88
        10
```

10.
```
       158
   22)3491
      2200
      1291
      1100
       191
       176
        15
```

11.
```
       229
   24)5503
      4800
       703
       480
       223
       216
         7
```

13.
```
       113
   23)2619
      2300
       319
       230
        89
        69
        20
```

14.
```
       121
   32)3901
      3200
       701
       640
        61
        32
        29
```

15.
```
       313
   14)4393
      4200
       193
       140
        53
        42
        11
```

```
          145              2.         108          3.          121
      38)5529                     44)4779                  25)3042
         3800                        4400                     2500
         1729                         379                      542
         1520                         352                      500
          209                          27                       42
          190                                                   25
           19                                                   17

                            6.         176
   5.         298                   12)2116          7.          150
      16)4773                         1200                  32)4812
         3200                          916                     3200
         1573                          840                     1612
         1440                           76                     1600
          133                           72                       12
          128                            4
            5

                           10.         168          11.         241
                               27)4544                  24)5797
   9.         273                     2700                     4800
      13)3558                         1844                      997
         2600                         1620                      960
          958                          224                       37
          910                          216                       24
           48                            8                       13
           39
            9

                           14.         331          15.         134
                               18)5959                  42)5661
  13.         140                     5400                     4200
      32)4511                          559                     1461
         3200                          540                     1260
         1311                           19                      201
         1280                           18                      168
           31                            1                       33
```

```
          348              2.         307          3.          158
      12)4179                     16)4923                  35)5540
         3600                        4800                     3500
          579                         123                     2040
          480                         112                     1750
           99                          11                      290
           96                                                  280
            3                                                   10

                            6.         229
   5.         148                   23)5278          7.          160
      15)2227                         4600                  27)4334
         1500                          678                     2700
          727                          460                     1634
          600                          218                     1620
          127                          207                       14
          120                           11
            7

                           10.         198          11.         151
                               18)3577                  41)6192
   9.         103                     1800                     4100
      39)4051                         1777                     2092
         3900                         1620                     2050
          151                          157                       42
          117                          144                       41
           34                           13                        1

  13.         312           14.         315          15.         119
      19)5943                   22)6935                  29)3456
         5700                         6600                     2900
          243                          335                      556
          190                          220                      290
           53                          115                      266
           38                          110                      261
           15                            5                        5
```

```
          113              2.         183          3.          494
      55)6237                     37)6805                  12)5929
         5500                        3700                     4800
          737                        3105                     1129
          550                        2960                     1080
          187                         145                       49
          165                         111                       48
           22                          34                        1

   5.         153           6.         241          7.          209
      33)5070                   28)6763                  23)4824
         3300                        5600                     4600
         1770                        1163                      224
         1650                        1120                      207
          120                          43                       17
           99                          28
           21                          15

                                                    11.         179
                                                        32)5737
   9.         127          10.         143                     3200
      26)3319                   26)3723                     2537
         2600                        2600                     2240
          719                        1123                      297
          520                        1040                      288
          199                          83                        9
          182                          78
           17                           5

                                                    15.         150
                                                        44)6611
  13.         263          14.         132                     4400
      11)2896                   42)5565                     2211
         2200                        4200                     2200
          696                        1365                       11
          660                        1260
           36                         105
           33                          84
            3                          21
```

```
          427              2.         265          3.          169
      13)5562                     12)3189                  31)5254
         5200                        2400                     3100
          362                         789                     2154
          260                         720                     1860
          102                          69                      294
           91                          60                      279
           11                           9                       15

   5.         141           6.         122          7.          258
      25)3528                   46)5635                  19)4920
         2500                        4600                     3800
         1028                        1035                     1120
         1000                         920                      950
           28                         115                      170
           25                          92                      152
            3                          23                       18

   9.         176          10.         241          11.         135
      35)6174                   16)3857                  39)5280
         3500                        3200                     3900
         2674                         657                     1380
         2450                         640                     1170
          224                          17                      210
          210                          16                      195
           14                           1                       15

  13.         135          14.         237          15.         234
      41)5542                   23)5472                  28)6555
         4100                        4600                     5600
         1442                         872                      955
         1230                         690                      840
          212                         182                      115
          205                         161                      112
            7                          21                        3
```

Change each fraction to simplest form.

1. $\dfrac{2}{4} = \dfrac{1}{2}$ 2. $\dfrac{4}{12} = \dfrac{1}{3}$ 3. $\dfrac{3}{15} = \dfrac{1}{5}$

4. $\dfrac{2}{6} = \dfrac{1}{3}$ 5. $\dfrac{8}{12} = \dfrac{2}{3}$ 6. $\dfrac{4}{12} = \dfrac{1}{3}$

7. $\dfrac{3}{9} = \dfrac{1}{3}$ 8. $\dfrac{4}{6} = \dfrac{2}{3}$ 9. $\dfrac{6}{9} = \dfrac{2}{3}$

10. $\dfrac{2}{8} = \dfrac{1}{4}$ 11. $\dfrac{2}{12} = \dfrac{1}{6}$ 12. $\dfrac{10}{12} = \dfrac{5}{6}$

13. $\dfrac{4}{8} = \dfrac{1}{2}$ 14. $\dfrac{6}{9} = \dfrac{2}{3}$ 15. $\dfrac{5}{15} = \dfrac{1}{3}$

16. $\dfrac{6}{8} = \dfrac{3}{4}$ 17. $\dfrac{6}{12} = \dfrac{1}{2}$ 18. $\dfrac{12}{18} = \dfrac{2}{3}$

19. $\dfrac{2}{2} = 1$ 20. $\dfrac{3}{9} = \dfrac{1}{3}$ 21. $\dfrac{8}{12} = \dfrac{2}{3}$

22. $\dfrac{6}{12} = \dfrac{1}{2}$ 23. $\dfrac{10}{12} = \dfrac{5}{6}$ 24. $\dfrac{9}{12} = \dfrac{3}{4}$

25. $\dfrac{3}{6} = \dfrac{1}{2}$ 26. $\dfrac{9}{9} = 1$ 27. $\dfrac{15}{18} = \dfrac{5}{6}$

Change each fraction to simplest form.

1. $\dfrac{5}{10} = \dfrac{1}{2}$ 2. $\dfrac{9}{12} = \dfrac{3}{4}$ 3. $\dfrac{12}{24} = \dfrac{1}{2}$

4. $\dfrac{4}{6} = \dfrac{2}{3}$ 5. $\dfrac{6}{10} = \dfrac{3}{5}$ 6. $\dfrac{2}{18} = \dfrac{1}{9}$

7. $\dfrac{8}{12} = \dfrac{2}{3}$ 8. $\dfrac{4}{12} = \dfrac{1}{3}$ 9. $\dfrac{15}{18} = \dfrac{5}{6}$

10. $\dfrac{6}{10} = \dfrac{3}{5}$ 11. $\dfrac{4}{10} = \dfrac{2}{5}$ 12. $\dfrac{4}{24} = \dfrac{1}{6}$

13. $\dfrac{6}{8} = \dfrac{3}{4}$ 14. $\dfrac{5}{15} = \dfrac{1}{3}$ 15. $\dfrac{10}{18} = \dfrac{5}{9}$

16. $\dfrac{10}{15} = \dfrac{2}{3}$ 17. $\dfrac{3}{18} = \dfrac{1}{6}$ 18. $\dfrac{17}{34} = \dfrac{1}{2}$

19. $\dfrac{3}{12} = \dfrac{1}{4}$ 20. $\dfrac{2}{6} = \dfrac{1}{3}$ 21. $\dfrac{6}{24} = \dfrac{1}{4}$

22. $\dfrac{2}{10} = \dfrac{1}{5}$ 23. $\dfrac{6}{12} = \dfrac{1}{2}$ 24. $\dfrac{7}{28} = \dfrac{1}{4}$

25. $\dfrac{12}{15} = \dfrac{4}{5}$ 26. $\dfrac{12}{20} = \dfrac{3}{5}$ 27. $\dfrac{15}{30} = \dfrac{1}{2}$

Change each fraction to simplest form.

1. $\dfrac{3}{12} = \dfrac{1}{4}$ 2. $\dfrac{4}{8} = \dfrac{1}{2}$ 3. $\dfrac{8}{12} = \dfrac{2}{3}$

4. $\dfrac{6}{8} = \dfrac{3}{4}$ 5. $\dfrac{2}{6} = \dfrac{1}{3}$ 6. $\dfrac{2}{8} = \dfrac{1}{4}$

7. $\dfrac{4}{6} = \dfrac{2}{3}$ 8. $\dfrac{4}{10} = \dfrac{2}{5}$ 9. $\dfrac{8}{10} = \dfrac{4}{5}$

10. $\dfrac{6}{10} = \dfrac{3}{5}$ 11. $\dfrac{4}{12} = \dfrac{1}{3}$ 12. $\dfrac{12}{15} = \dfrac{4}{5}$

13. $\dfrac{7}{14} = \dfrac{1}{2}$ 14. $\dfrac{5}{15} = \dfrac{1}{3}$ 15. $\dfrac{9}{18} = \dfrac{1}{2}$

16. $\dfrac{10}{15} = \dfrac{2}{3}$ 17. $\dfrac{15}{18} = \dfrac{5}{6}$ 18. $\dfrac{18}{24} = \dfrac{3}{4}$

19. $\dfrac{6}{12} = \dfrac{1}{2}$ 20. $\dfrac{16}{24} = \dfrac{2}{3}$ 21. $\dfrac{22}{24} = \dfrac{11}{12}$

22. $\dfrac{12}{18} = \dfrac{2}{3}$ 23. $\dfrac{20}{24} = \dfrac{5}{6}$ 24. $\dfrac{10}{12} = \dfrac{5}{6}$

25. $\dfrac{10}{24} = \dfrac{5}{12}$ 26. $\dfrac{10}{25} = \dfrac{2}{5}$ 27. $\dfrac{20}{25} = \dfrac{4}{5}$

Change each fraction to simplest form.

1. $\dfrac{4}{10} = \dfrac{2}{5}$ 2. $\dfrac{6}{12} = \dfrac{1}{2}$ 3. $\dfrac{3}{15} = \dfrac{1}{5}$

4. $\dfrac{10}{12} = \dfrac{5}{6}$ 5. $\dfrac{6}{10} = \dfrac{3}{5}$ 6. $\dfrac{8}{12} = \dfrac{2}{3}$

7. $\dfrac{10}{15} = \dfrac{2}{3}$ 8. $\dfrac{5}{15} = \dfrac{1}{3}$ 9. $\dfrac{8}{10} = \dfrac{4}{5}$

10. $\dfrac{6}{18} = \dfrac{1}{3}$ 11. $\dfrac{8}{18} = \dfrac{4}{9}$ 12. $\dfrac{10}{18} = \dfrac{5}{9}$

13. $\dfrac{12}{20} = \dfrac{3}{5}$ 14. $\dfrac{2}{20} = \dfrac{1}{10}$ 15. $\dfrac{5}{20} = \dfrac{1}{4}$

16. $\dfrac{16}{20} = \dfrac{4}{5}$ 17. $\dfrac{18}{20} = \dfrac{9}{10}$ 18. $\dfrac{4}{24} = \dfrac{1}{6}$

19. $\dfrac{18}{24} = \dfrac{3}{4}$ 20. $\dfrac{20}{24} = \dfrac{5}{6}$ 21. $\dfrac{22}{24} = \dfrac{11}{12}$

22. $\dfrac{15}{30} = \dfrac{1}{2}$ 23. $\dfrac{6}{30} = \dfrac{1}{5}$ 24. $\dfrac{8}{30} = \dfrac{4}{15}$

25. $\dfrac{4}{32} = \dfrac{1}{8}$ 26. $\dfrac{6}{32} = \dfrac{3}{16}$ 27. $\dfrac{20}{32} = \dfrac{5}{8}$

Fill in the blank with $<$, $>$, or $=$ to make each statement true.

1. $\dfrac{1}{2}$ $<$ 1 2. $\dfrac{3}{3}$ $=$ 1 3. $\dfrac{1}{2}$ $=$ $\dfrac{6}{12}$

4. $\dfrac{1}{4}$ $<$ $\dfrac{2}{3}$ 5. $\dfrac{5}{4}$ $>$ $\dfrac{2}{3}$ 6. $\dfrac{4}{3}$ $>$ $\dfrac{2}{3}$

7. $\dfrac{1}{3}$ $>$ $\dfrac{1}{6}$ 8. $\dfrac{2}{6}$ $=$ $\dfrac{1}{3}$ 9. $\dfrac{8}{9}$ $>$ $\dfrac{4}{6}$

10. $\dfrac{2}{5}$ $<$ $\dfrac{3}{5}$ 11. $\dfrac{3}{7}$ $<$ $\dfrac{2}{3}$ 12. $\dfrac{10}{3}$ $>$ 1

13. 1 $>$ $\dfrac{3}{4}$ 14. $\dfrac{6}{10}$ $=$ $\dfrac{3}{5}$ 15. $\dfrac{25}{20}$ $=$ $\dfrac{5}{4}$

16. $\dfrac{5}{6}$ $>$ $\dfrac{1}{2}$ 17. $\dfrac{6}{5}$ $>$ 1 18. $\dfrac{11}{33}$ $=$ $\dfrac{2}{6}$

19. $\dfrac{1}{3}$ $>$ $\dfrac{1}{5}$ 20. $\dfrac{12}{4}$ $>$ 1 21. $\dfrac{2}{3}$ $<$ $\dfrac{9}{12}$

22. $\dfrac{2}{5}$ $<$ $\dfrac{1}{2}$ 23. $\dfrac{5}{8}$ $>$ $\dfrac{1}{2}$ 24. $\dfrac{5}{6}$ $>$ $\dfrac{5}{7}$

25. $\dfrac{2}{7}$ $<$ $\dfrac{3}{7}$ 26. $\dfrac{9}{5}$ $>$ $\dfrac{6}{7}$ 27. $\dfrac{21}{28}$ $=$ $\dfrac{12}{16}$

Fill in the blank with $<$, $>$, or $=$ to make each statement true.

1. $\dfrac{2}{3}$ $>$ $\dfrac{5}{9}$ 2. $\dfrac{2}{2}$ $<$ $\dfrac{4}{3}$ 3. $\dfrac{1}{4}$ $>$ $\dfrac{1}{8}$

4. $\dfrac{1}{2}$ $<$ $\dfrac{3}{4}$ 5. $\dfrac{1}{3}$ $<$ $\dfrac{1}{2}$ 6. $\dfrac{3}{4}$ $=$ $\dfrac{9}{12}$

7. $\dfrac{3}{4}$ $>$ $\dfrac{1}{3}$ 8. 1 $<$ $\dfrac{4}{3}$ 9. $\dfrac{3}{3}$ $=$ $\dfrac{7}{7}$

10. $\dfrac{2}{5}$ $>$ $\dfrac{2}{6}$ 11. $\dfrac{4}{7}$ $>$ $\dfrac{1}{2}$ 12. 1 $<$ $\dfrac{7}{6}$

13. $\dfrac{3}{7}$ $<$ $\dfrac{4}{8}$ 14. $\dfrac{2}{8}$ $<$ $\dfrac{1}{3}$ 15. $\dfrac{5}{15}$ $=$ $\dfrac{3}{9}$

16. $\dfrac{5}{8}$ $>$ $\dfrac{1}{2}$ 17. $\dfrac{10}{12}$ $=$ $\dfrac{20}{24}$ 18. $\dfrac{8}{12}$ $>$ $\dfrac{10}{20}$

19. $\dfrac{6}{12}$ $<$ $\dfrac{2}{3}$ 20. $\dfrac{8}{7}$ $>$ $\dfrac{5}{5}$ 21. $\dfrac{9}{12}$ $>$ $\dfrac{4}{6}$

22. $\dfrac{9}{18}$ $=$ $\dfrac{11}{22}$ 23. $\dfrac{24}{8}$ $>$ 1 24. $\dfrac{21}{28}$ $=$ $\dfrac{27}{36}$

25. $\dfrac{15}{30}$ $=$ $\dfrac{16}{32}$ 26. $\dfrac{7}{21}$ $=$ $\dfrac{6}{18}$ 27. $\dfrac{6}{5}$ $=$ $\dfrac{18}{15}$

Fill in the blank with $<$, $>$, or $=$ to make each statement true.

1. $\dfrac{4}{5}$ $>$ $\dfrac{3}{4}$ 2. $\dfrac{2}{6}$ $>$ $\dfrac{1}{4}$ 3. $\dfrac{5}{5}$ $>$ $\dfrac{5}{6}$

4. $\dfrac{1}{6}$ $<$ $\dfrac{3}{3}$ 5. $\dfrac{2}{5}$ $=$ $\dfrac{6}{15}$ 6. $\dfrac{8}{9}$ $=$ $\dfrac{16}{18}$

7. $\dfrac{2}{7}$ $>$ $\dfrac{1}{6}$ 8. $\dfrac{3}{7}$ $<$ $\dfrac{2}{3}$ 9. $\dfrac{4}{5}$ $>$ $\dfrac{2}{3}$

10. $\dfrac{14}{16}$ $>$ $\dfrac{5}{6}$ 11. $\dfrac{4}{6}$ $=$ $\dfrac{10}{15}$ 12. $\dfrac{7}{10}$ $>$ $\dfrac{1}{2}$

13. $\dfrac{17}{5}$ $>$ $\dfrac{15}{7}$ 14. $\dfrac{7}{21}$ $<$ $\dfrac{2}{3}$ 15. $\dfrac{9}{15}$ $>$ $\dfrac{5}{10}$

16. $\dfrac{11}{21}$ $<$ $\dfrac{13}{21}$ 17. $\dfrac{4}{3}$ $=$ $\dfrac{16}{12}$ 18. $\dfrac{9}{5}$ $<$ $\dfrac{19}{6}$

19. $\dfrac{13}{6}$ $>$ $\dfrac{11}{6}$ 20. $\dfrac{7}{2}$ $>$ $\dfrac{19}{7}$ 21. $\dfrac{24}{4}$ $>$ $\dfrac{36}{9}$

22. $\dfrac{9}{18}$ $=$ $\dfrac{17}{34}$ 23. $\dfrac{28}{35}$ $=$ $\dfrac{16}{20}$ 24. $\dfrac{8}{9}$ $=$ $\dfrac{24}{27}$

25. $\dfrac{12}{30}$ $=$ $\dfrac{4}{10}$ 26. $\dfrac{4}{10}$ $<$ $\dfrac{9}{15}$ 27. $\dfrac{7}{21}$ $=$ $\dfrac{16}{48}$

Find equivalent fractions.

1. $\dfrac{1}{2}$ $=$ $\dfrac{2}{4}$ $=$ $\dfrac{3}{6}$ $=$ $\dfrac{5}{10}$ $=$ $\dfrac{6}{12}$ $=$ $\dfrac{9}{18}$

2. $\dfrac{1}{3}$ $=$ $\dfrac{2}{6}$ $=$ $\dfrac{4}{12}$ $=$ $\dfrac{5}{15}$ $=$ $\dfrac{7}{21}$ $=$ $\dfrac{10}{30}$

3. $\dfrac{1}{5}$ $=$ $\dfrac{3}{15}$ $=$ $\dfrac{4}{20}$ $=$ $\dfrac{6}{30}$ $=$ $\dfrac{7}{35}$ $=$ $\dfrac{10}{50}$

4. $\dfrac{2}{3}$ $=$ $\dfrac{4}{6}$ $=$ $\dfrac{6}{9}$ $=$ $\dfrac{10}{15}$ $=$ $\dfrac{12}{18}$ $=$ $\dfrac{16}{24}$

5. $\dfrac{2}{5}$ $=$ $\dfrac{6}{15}$ $=$ $\dfrac{10}{25}$ $=$ $\dfrac{12}{30}$ $=$ $\dfrac{16}{40}$ $=$ $\dfrac{22}{55}$

6. $\dfrac{1}{6}$ $=$ $\dfrac{2}{12}$ $=$ $\dfrac{4}{24}$ $=$ $\dfrac{7}{42}$ $=$ $\dfrac{8}{48}$ $=$ $\dfrac{9}{54}$

7. $\dfrac{3}{4}$ $=$ $\dfrac{6}{8}$ $=$ $\dfrac{9}{12}$ $=$ $\dfrac{15}{20}$ $=$ $\dfrac{21}{28}$ $=$ $\dfrac{27}{36}$

8. $\dfrac{3}{5}$ $=$ $\dfrac{9}{15}$ $=$ $\dfrac{12}{20}$ $=$ $\dfrac{15}{25}$ $=$ $\dfrac{18}{30}$ $=$ $\dfrac{33}{55}$

9. $\dfrac{5}{6}$ $=$ $\dfrac{10}{12}$ $=$ $\dfrac{20}{24}$ $=$ $\dfrac{25}{30}$ $=$ $\dfrac{30}{36}$ $=$ $\dfrac{45}{54}$

Find equivalent fractions.

1. $\dfrac{2}{3} = \dfrac{4}{6} = \dfrac{8}{12} = \dfrac{10}{15} = \dfrac{14}{21} = \dfrac{16}{24}$

2. $\dfrac{3}{4} = \dfrac{9}{12} = \dfrac{12}{16} = \dfrac{18}{24} = \dfrac{21}{28} = \dfrac{27}{36}$

3. $\dfrac{3}{5} = \dfrac{9}{15} = \dfrac{12}{20} = \dfrac{15}{25} = \dfrac{21}{35} = \dfrac{24}{40}$

4. $\dfrac{2}{7} = \dfrac{4}{14} = \dfrac{6}{21} = \dfrac{10}{35} = \dfrac{12}{42} = \dfrac{16}{56}$

5. $\dfrac{4}{5} = \dfrac{8}{10} = \dfrac{16}{20} = \dfrac{20}{25} = \dfrac{24}{30} = \dfrac{40}{50}$

6. $\dfrac{3}{7} = \dfrac{9}{21} = \dfrac{15}{35} = \dfrac{18}{42} = \dfrac{21}{49} = \dfrac{27}{63}$

7. $\dfrac{4}{9} = \dfrac{12}{27} = \dfrac{20}{45} = \dfrac{28}{63} = \dfrac{32}{72} = \dfrac{36}{81}$

8. $\dfrac{5}{7} = \dfrac{10}{14} = \dfrac{25}{35} = \dfrac{30}{42} = \dfrac{35}{49} = \dfrac{40}{56}$

9. $\dfrac{5}{9} = \dfrac{20}{36} = \dfrac{25}{45} = \dfrac{30}{54} = \dfrac{35}{63} = \dfrac{40}{72}$

Find equivalent fractions.

1. $\dfrac{5}{7} = \dfrac{10}{14} = \dfrac{15}{21} = \dfrac{35}{49} = \dfrac{40}{56} = \dfrac{45}{63}$

2. $\dfrac{2}{9} = \dfrac{4}{18} = \dfrac{6}{27} = \dfrac{12}{54} = \dfrac{14}{63} = \dfrac{18}{81}$

3. $\dfrac{5}{6} = \dfrac{15}{18} = \dfrac{20}{24} = \dfrac{30}{36} = \dfrac{35}{42} = \dfrac{55}{66}$

4. $\dfrac{6}{7} = \dfrac{18}{21} = \dfrac{24}{28} = \dfrac{42}{49} = \dfrac{48}{56} = \dfrac{72}{84}$

5. $\dfrac{4}{9} = \dfrac{20}{45} = \dfrac{24}{54} = \dfrac{36}{81} = \dfrac{40}{90} = \dfrac{48}{108}$

6. $\dfrac{2}{7} = \dfrac{10}{35} = \dfrac{12}{42} = \dfrac{16}{56} = \dfrac{18}{63} = \dfrac{26}{91}$

7. $\dfrac{3}{10} = \dfrac{9}{30} = \dfrac{12}{40} = \dfrac{21}{70} = \dfrac{24}{80} = \dfrac{45}{150}$

8. $\dfrac{7}{9} = \dfrac{21}{27} = \dfrac{28}{36} = \dfrac{49}{63} = \dfrac{56}{72} = \dfrac{98}{126}$

9. $\dfrac{9}{11} = \dfrac{18}{22} = \dfrac{36}{44} = \dfrac{45}{55} = \dfrac{81}{99} = \dfrac{99}{121}$

Add fractions and write the answers in simplest form.

1. $\dfrac{1}{6} + \dfrac{1}{6} = \dfrac{2}{6} = \dfrac{1}{3}$ 2. $\dfrac{1}{6} + \dfrac{2}{6} = \dfrac{3}{6} = \dfrac{1}{2}$

3. $\dfrac{1}{4} + \dfrac{1}{4} = \dfrac{2}{4} = \dfrac{1}{2}$ 4. $\dfrac{2}{4} + \dfrac{6}{4} = \dfrac{8}{4} = 2$

5. $\dfrac{2}{10} + \dfrac{3}{10} = \dfrac{5}{10} = \dfrac{1}{2}$ 6. $\dfrac{1}{10} + \dfrac{5}{10} = \dfrac{6}{10} = \dfrac{3}{5}$

7. $\dfrac{2}{8} + \dfrac{4}{8} = \dfrac{6}{8} = \dfrac{3}{4}$ 8. $\dfrac{3}{5} + \dfrac{2}{5} = \dfrac{5}{5} = 1$

9. $\dfrac{1}{5} + \dfrac{2}{5} = \dfrac{3}{5}$ 10. $\dfrac{1}{8} + \dfrac{3}{8} = \dfrac{4}{8} = \dfrac{1}{2}$

11. $\dfrac{1}{9} + \dfrac{2}{9} = \dfrac{3}{9} = \dfrac{1}{3}$ 12. $\dfrac{3}{12} + \dfrac{3}{12} = \dfrac{6}{12} = \dfrac{1}{2}$

13. $\dfrac{2}{12} + \dfrac{1}{12} = \dfrac{3}{12} = \dfrac{1}{4}$ 14. $\dfrac{2}{9} + \dfrac{4}{9} = \dfrac{6}{9} = \dfrac{2}{3}$

15. $\dfrac{3}{10} + \dfrac{5}{10} = \dfrac{8}{10} = \dfrac{4}{5}$ 16. $\dfrac{5}{10} + \dfrac{1}{10} = \dfrac{6}{10} = \dfrac{3}{5}$

17. $\dfrac{3}{12} + \dfrac{2}{12} = \dfrac{5}{12}$ 18. $\dfrac{4}{12} + \dfrac{6}{12} = \dfrac{10}{12} = \dfrac{5}{6}$

Add fractions and write the answers in simplest form.

1. $\dfrac{3}{4} + \dfrac{3}{4} = \dfrac{6}{4} = \dfrac{3}{2}$ 2. $\dfrac{1}{8} + \dfrac{3}{8} = \dfrac{4}{8} = \dfrac{1}{2}$

3. $\dfrac{2}{5} + \dfrac{3}{5} = \dfrac{5}{5} = 1$ 4. $\dfrac{1}{4} + \dfrac{1}{4} = \dfrac{2}{4} = \dfrac{1}{2}$

5. $\dfrac{3}{8} + \dfrac{3}{8} = \dfrac{6}{8} = \dfrac{3}{4}$ 6. $\dfrac{3}{10} + \dfrac{2}{10} = \dfrac{5}{10} = \dfrac{1}{2}$

7. $\dfrac{1}{10} + \dfrac{1}{10} = \dfrac{2}{10} = \dfrac{1}{5}$ 8. $\dfrac{4}{9} + \dfrac{2}{9} = \dfrac{6}{9} = \dfrac{2}{3}$

9. $\dfrac{2}{9} + \dfrac{1}{9} = \dfrac{3}{9} = \dfrac{1}{3}$ 10. $\dfrac{1}{12} + \dfrac{1}{12} = \dfrac{2}{12} = \dfrac{1}{6}$

11. $\dfrac{2}{12} + \dfrac{2}{12} = \dfrac{4}{12} = \dfrac{1}{3}$ 12. $\dfrac{7}{12} + \dfrac{9}{12} = \dfrac{16}{12} = \dfrac{4}{3}$

13. $\dfrac{7}{12} + \dfrac{3}{12} = \dfrac{10}{12} = \dfrac{5}{6}$ 14. $\dfrac{4}{15} + \dfrac{6}{15} = \dfrac{10}{15} = \dfrac{2}{3}$

15. $\dfrac{3}{15} + \dfrac{2}{15} = \dfrac{5}{15} = \dfrac{1}{3}$ 16. $\dfrac{8}{15} + \dfrac{7}{15} = \dfrac{15}{15} = 1$

17. $\dfrac{1}{14} + \dfrac{1}{14} = \dfrac{2}{14} = \dfrac{1}{7}$ 18. $\dfrac{5}{14} + \dfrac{2}{14} = \dfrac{7}{14} = \dfrac{1}{2}$

Lesson 4-3 Adding fractions with the same denominators

Name: _____

Add fractions and write the answers in simplest form.

1. $\frac{1}{2} + \frac{3}{2} = \frac{4}{2} = 2$
2. $\frac{1}{4} + \frac{7}{4} = \frac{8}{4} = 2$
3. $\frac{2}{4} + \frac{5}{4} = \underline{\quad} = \frac{7}{4}$
4. $\frac{1}{6} + \frac{2}{6} = \frac{3}{6} = \frac{1}{2}$
5. $\frac{3}{6} + \frac{1}{6} = \frac{4}{6} = \frac{2}{3}$
6. $\frac{5}{8} + \frac{7}{8} = \frac{12}{8} = \frac{3}{2}$
7. $\frac{6}{8} + \frac{4}{8} = \frac{10}{8} = \frac{5}{4}$
8. $\frac{7}{9} + \frac{8}{9} = \frac{15}{9} = \frac{5}{3}$
9. $\frac{7}{9} + \frac{5}{9} = \frac{12}{9} = \frac{4}{3}$
10. $\frac{6}{12} + \frac{8}{12} = \frac{14}{12} = \frac{7}{6}$
11. $\frac{4}{12} + \frac{4}{12} = \frac{8}{12} = \frac{2}{3}$
12. $\frac{10}{12} + \frac{8}{12} = \frac{18}{12} = \frac{3}{2}$
13. $\frac{5}{10} + \frac{7}{10} = \frac{12}{10} = \frac{6}{5}$
14. $\frac{8}{10} + \frac{12}{10} = \frac{20}{10} = 2$
15. $\frac{6}{15} + \frac{9}{15} = \frac{15}{15} = 1$
16. $\frac{6}{15} + \frac{6}{15} = \frac{12}{15} = \frac{4}{5}$
17. $\frac{10}{12} + \frac{10}{12} = \frac{20}{12} = \frac{5}{3}$
18. $\frac{12}{8} + \frac{6}{8} = \frac{18}{8} = \frac{9}{4}$

Lesson 4-4 Adding fractions with the same denominators

Name: _____

Add fractions and write the answers in simplest form.

1. $\frac{1}{2} + \frac{1}{2} = \frac{2}{2} = 1$
2. $1\frac{1}{3} + \frac{1}{3} = 1\frac{2}{3}$
3. $1\frac{1}{3} + 1\frac{2}{3} = 2\frac{3}{3} = 3$
4. $\frac{1}{4} + 2\frac{3}{4} = 2\frac{4}{4} = 3$
5. $2\frac{1}{4} + \frac{1}{4} = 2\frac{2}{4} = 2\frac{1}{2}$
6. $3\frac{1}{6} + 1\frac{1}{6} = 4\frac{2}{6} = 4\frac{1}{3}$
7. $2\frac{1}{6} + 2\frac{2}{6} = 4\frac{3}{6} = 4\frac{1}{2}$
8. $1\frac{1}{8} + \frac{3}{8} = 1\frac{4}{8} = 1\frac{1}{2}$
9. $\frac{1}{8} + 4\frac{3}{8} = 4\frac{4}{8} = 4\frac{1}{2}$
10. $4\frac{2}{8} + 3\frac{2}{8} = 7\frac{4}{8} = 7\frac{1}{2}$
11. $2\frac{3}{8} + 3\frac{3}{8} = 5\frac{6}{8} = 5\frac{3}{4}$
12. $2\frac{1}{9} + 4\frac{5}{9} = 6\frac{6}{9} = 6\frac{2}{3}$
13. $3\frac{2}{9} + 3\frac{1}{9} = 6\frac{3}{9} = 6\frac{1}{3}$
14. $4\frac{3}{10} + \frac{1}{10} = 4\frac{4}{10} = 4\frac{2}{5}$
15. $5\frac{5}{10} + \frac{5}{10} = 5\frac{10}{10} = 6$
16. $4\frac{1}{12} + 4\frac{1}{12} = 8\frac{2}{12} = 8\frac{1}{6}$
17. $5\frac{2}{12} + 4\frac{1}{12} = 9\frac{3}{12} = 9\frac{1}{4}$
18. $\frac{3}{15} + 3\frac{2}{15} = 3\frac{5}{15} = 3\frac{1}{3}$

Lesson 4-5 Adding fractions with the same denominators

Name: _____

Add fractions and write the answers in simplest form.

1. $2\frac{3}{6} + 1\frac{1}{6} = 3\frac{4}{6} = 3\frac{2}{3}$
2. $3\frac{1}{8} + 2\frac{3}{8} = 5\frac{4}{8} = 5\frac{1}{2}$
3. $1\frac{1}{8} + \frac{1}{8} = 1\frac{2}{8} = 1\frac{1}{4}$
4. $2\frac{1}{10} + 3\frac{3}{10} = 5\frac{4}{10} = 5\frac{2}{5}$
5. $3\frac{3}{10} + 2\frac{2}{10} = 5\frac{5}{10} = 5\frac{1}{2}$
6. $5\frac{2}{8} + \frac{4}{8} = 5\frac{6}{8} = 5\frac{3}{4}$
7. $5\frac{1}{12} + 4\frac{1}{12} = 9\frac{2}{12} = 9\frac{1}{6}$
8. $4\frac{1}{12} + 3\frac{2}{12} = 7\frac{3}{12} = 7\frac{1}{4}$
9. $\frac{3}{12} + 5\frac{3}{12} = 5\frac{6}{12} = 5\frac{1}{2}$
10. $6\frac{5}{12} + 4\frac{3}{12} = 10\frac{8}{12} = 10\frac{2}{3}$
11. $1\frac{2}{15} + 2\frac{1}{15} = 3\frac{3}{15} = 3\frac{1}{5}$
12. $\frac{2}{15} + 1\frac{3}{15} = 1\frac{5}{15} = 1\frac{1}{3}$
13. $4\frac{3}{18} + 2\frac{3}{18} = 6\frac{6}{18} = 6\frac{1}{3}$
14. $5\frac{5}{18} + 6\frac{4}{18} = 11\frac{9}{18} = 11\frac{1}{2}$
15. $\frac{3}{20} + 5\frac{7}{20} = 5\frac{10}{20} = 5\frac{1}{2}$
16. $2\frac{5}{16} + \frac{3}{16} = 2\frac{8}{16} = 2\frac{1}{2}$
17. $7\frac{1}{16} + 3\frac{1}{16} = 10\frac{2}{16} = 10\frac{1}{8}$
18. $4\frac{3}{20} + 7\frac{2}{20} = 11\frac{5}{20} = 11\frac{1}{4}$

Lesson 4-6 Subtracting fractions with the same denominators

Name: _____

Subtract fractions and write the answers in simplest form.

1. $\frac{2}{2} - \frac{1}{2} = \frac{1}{2}$
2. $\frac{3}{4} - \frac{2}{4} = \frac{1}{4}$
3. $\frac{3}{4} - \frac{1}{4} = \frac{2}{4} = \frac{1}{2}$
4. $\frac{5}{8} - \frac{1}{8} = \frac{4}{8} = \frac{1}{2}$
5. $\frac{4}{5} - \frac{2}{5} = \frac{2}{5}$
6. $\frac{7}{8} - \frac{3}{8} = \frac{4}{8} = \frac{1}{2}$
7. $\frac{3}{6} - \frac{1}{6} = \frac{2}{6} = \frac{1}{3}$
8. $\frac{8}{10} - \frac{4}{10} = \frac{4}{10} = \frac{2}{5}$
9. $\frac{9}{10} - \frac{4}{10} = \frac{5}{10} = \frac{1}{2}$
10. $\frac{7}{12} - \frac{3}{12} = \frac{4}{12} = \frac{1}{3}$
11. $\frac{5}{10} - \frac{3}{10} = \frac{2}{10} = \frac{1}{5}$
12. $\frac{11}{12} - \frac{1}{12} = \frac{10}{12} = \frac{5}{6}$
13. $\frac{12}{12} - \frac{2}{12} = \frac{10}{12} = \frac{5}{6}$
14. $\frac{10}{15} - \frac{3}{15} = \frac{7}{15}$
15. $\frac{8}{12} - \frac{5}{12} = \frac{3}{12} = \frac{1}{4}$
16. $\frac{13}{18} - \frac{4}{18} = \frac{9}{18} = \frac{1}{2}$
17. $\frac{12}{18} - \frac{10}{18} = \frac{2}{18} = \frac{1}{9}$
18. $\frac{14}{15} - \frac{4}{15} = \frac{10}{15} = \frac{2}{3}$

Lesson 4-7   Subtracting fractions with the same denominators

Subtract fractions and write the answers in simplest form.

1. $\dfrac{4}{6} - \dfrac{2}{6} = \dfrac{2}{6} = \dfrac{1}{3}$

2. $\dfrac{8}{9} - \dfrac{2}{9} = \dfrac{6}{9} = \dfrac{2}{3}$

3. $\dfrac{4}{7} - \dfrac{2}{7} = \dfrac{2}{7}$

4. $\dfrac{5}{8} - \dfrac{2}{8} = \dfrac{3}{8}$

5. $\dfrac{7}{9} - \dfrac{1}{9} = \dfrac{6}{9} = \dfrac{2}{3}$

6. $\dfrac{7}{10} - \dfrac{2}{10} = \dfrac{5}{10} = \dfrac{1}{2}$

7. $\dfrac{11}{10} - \dfrac{3}{10} = \dfrac{8}{10} = \dfrac{4}{5}$

8. $\dfrac{7}{12} - \dfrac{2}{12} = \dfrac{5}{12}$

9. $\dfrac{15}{12} - \dfrac{6}{12} = \dfrac{9}{12} = \dfrac{3}{4}$

10. $\dfrac{9}{15} - \dfrac{5}{15} = \dfrac{4}{15}$

11. $\dfrac{11}{15} - \dfrac{2}{15} = \dfrac{9}{15} = \dfrac{3}{5}$

12. $\dfrac{13}{14} - \dfrac{6}{14} = \dfrac{7}{14} = \dfrac{1}{2}$

13. $\dfrac{9}{14} - \dfrac{3}{14} = \dfrac{6}{14} = \dfrac{3}{7}$

14. $\dfrac{15}{16} - \dfrac{3}{16} = \dfrac{12}{16} = \dfrac{3}{4}$

15. $\dfrac{10}{11} - \dfrac{3}{11} = \dfrac{7}{11}$

16. $\dfrac{10}{13} - \dfrac{3}{13} = \dfrac{7}{13}$

17. $\dfrac{14}{20} - \dfrac{8}{20} = \dfrac{6}{20} = \dfrac{3}{10}$

18. $\dfrac{18}{20} - \dfrac{4}{20} = \dfrac{14}{20} = \dfrac{7}{10}$

172

Lesson 4-8   Subtracting fractions with the same denominators

Subtract fractions and write the answers in simplest form.

1. $\dfrac{6}{7} - \dfrac{2}{7} = \dfrac{4}{7}$

2. $\dfrac{3}{8} - \dfrac{2}{8} = \dfrac{1}{8}$

3. $\dfrac{7}{8} - \dfrac{5}{8} = \dfrac{2}{8} = \dfrac{1}{4}$

4. $\dfrac{7}{9} - \dfrac{1}{9} = \dfrac{6}{9} = \dfrac{2}{3}$

5. $\dfrac{5}{9} - \dfrac{2}{9} = \dfrac{3}{9} = \dfrac{1}{3}$

6. $\dfrac{9}{10} - \dfrac{3}{10} = \dfrac{6}{10} = \dfrac{3}{5}$

7. $\dfrac{7}{10} - \dfrac{2}{10} = \dfrac{5}{10} = \dfrac{1}{2}$

8. $\dfrac{11}{12} - \dfrac{8}{12} = \dfrac{3}{12} = \dfrac{1}{4}$

9. $\dfrac{11}{12} - \dfrac{2}{12} = \dfrac{9}{12} = \dfrac{3}{4}$

10. $\dfrac{9}{14} - \dfrac{2}{14} = \dfrac{7}{14} = \dfrac{1}{2}$

11. $\dfrac{8}{13} - \dfrac{1}{13} = \dfrac{7}{13}$

12. $\dfrac{14}{15} - \dfrac{2}{15} = \dfrac{12}{15} = \dfrac{4}{5}$

13. $\dfrac{6}{11} - \dfrac{3}{11} = \dfrac{3}{11}$

14. $\dfrac{11}{18} - \dfrac{5}{18} = \dfrac{6}{18} = \dfrac{1}{3}$

15. $\dfrac{17}{18} - \dfrac{2}{18} = \dfrac{15}{18} = \dfrac{5}{6}$

16. $\dfrac{7}{17} - \dfrac{4}{17} = \dfrac{3}{17}$

17. $\dfrac{13}{16} - \dfrac{3}{16} = \dfrac{10}{16} = \dfrac{5}{8}$

18. $\dfrac{18}{20} - \dfrac{3}{20} = \dfrac{15}{20} = \dfrac{3}{4}$

173

Lesson 4-9   Subtracting fractions with the same denominators

Subtract fractions and write the answers in simplest form.

1. $1\dfrac{2}{3} - \dfrac{1}{3} = 1\dfrac{1}{3}$

2. $1\dfrac{2}{5} - \dfrac{1}{5} = 1\dfrac{1}{5}$

3. $3\dfrac{3}{4} - 1\dfrac{1}{4} = 2\dfrac{2}{4} = 2\dfrac{1}{2}$

4. $3\dfrac{6}{8} - 1\dfrac{4}{8} = 2\dfrac{2}{8} = 2\dfrac{1}{4}$

5. $2\dfrac{3}{6} - 2\dfrac{1}{6} = \dfrac{2}{6} = \dfrac{1}{3}$

6. $4\dfrac{7}{9} - 2\dfrac{1}{9} = 2\dfrac{6}{9} = 2\dfrac{2}{3}$

7. $5\dfrac{2}{5} - \dfrac{1}{5} = 5\dfrac{1}{5}$

8. $5\dfrac{7}{10} - 4\dfrac{5}{10} = 1\dfrac{2}{10} = 1\dfrac{1}{5}$

9. $4\dfrac{4}{8} - 2\dfrac{2}{8} = 2\dfrac{2}{8} = 2\dfrac{1}{4}$

10. $3\dfrac{7}{12} - 3\dfrac{3}{12} = \dfrac{4}{12} = \dfrac{1}{3}$

11. $6\dfrac{6}{10} - \dfrac{2}{10} = 6\dfrac{4}{10} = 6\dfrac{2}{5}$

12. $7\dfrac{11}{12} - 2\dfrac{1}{12} = 5\dfrac{10}{12} = 5\dfrac{5}{6}$

13. $7\dfrac{9}{12} - 2\dfrac{3}{12} = 5\dfrac{6}{12} = 5\dfrac{1}{2}$

14. $6\dfrac{5}{13} - \dfrac{2}{13} = 6\dfrac{3}{13}$

15. $4\dfrac{9}{15} - 1\dfrac{3}{15} = 3\dfrac{6}{15} = 3\dfrac{2}{5}$

16. $4\dfrac{7}{15} - 1\dfrac{4}{15} = 3\dfrac{3}{15} = 3\dfrac{1}{5}$

17. $6\dfrac{12}{18} - 3\dfrac{2}{18} = 3\dfrac{10}{18} = 3\dfrac{5}{9}$

18. $7\dfrac{10}{18} - 3\dfrac{6}{18} = 4\dfrac{4}{18} = 4\dfrac{2}{9}$

174

Lesson 4-10   Subtracting fractions with the same denominators

Subtract fractions and write the answers in simplest form.

1. $3\dfrac{5}{6} - \dfrac{2}{6} = 3\dfrac{3}{6} = 3\dfrac{1}{2}$

2. $4\dfrac{4}{5} - \dfrac{2}{5} = 4\dfrac{2}{5}$

3. $2\dfrac{5}{8} - 1\dfrac{4}{8} = 1\dfrac{1}{8}$

4. $3\dfrac{3}{4} - 1\dfrac{1}{4} = 2\dfrac{2}{4} = 2\dfrac{1}{2}$

5. $4\dfrac{7}{9} - 2\dfrac{4}{9} = 2\dfrac{3}{9} = 2\dfrac{1}{3}$

6. $5\dfrac{5}{6} - 2\dfrac{1}{6} = 3\dfrac{4}{6} = 3\dfrac{2}{3}$

7. $3\dfrac{8}{10} - 1\dfrac{4}{10} = 2\dfrac{4}{10} = 2\dfrac{2}{5}$

8. $7\dfrac{7}{11} - 4\dfrac{3}{11} = 3\dfrac{4}{11}$

9. $6\dfrac{10}{12} - 3\dfrac{5}{12} = 3\dfrac{5}{12}$

10. $5\dfrac{8}{12} - 3\dfrac{4}{12} = 2\dfrac{4}{12} = 2\dfrac{1}{3}$

11. $5\dfrac{8}{13} - 2\dfrac{3}{18} = 3\dfrac{5}{13}$

12. $4\dfrac{9}{13} - \dfrac{3}{13} = 4\dfrac{6}{13}$

13. $7\dfrac{8}{15} - 4\dfrac{5}{15} = 3\dfrac{3}{15} = 3\dfrac{1}{5}$

14. $6\dfrac{14}{15} - 5\dfrac{5}{15} = 1\dfrac{9}{15} = 1\dfrac{3}{5}$

15. $6\dfrac{9}{16} - 3\dfrac{5}{16} = 3\dfrac{4}{16} = 3\dfrac{1}{4}$

16. $4\dfrac{11}{16} - \dfrac{1}{16} = 4\dfrac{10}{16} = 4\dfrac{5}{8}$

17. $5\dfrac{12}{18} - \dfrac{9}{18} = 5\dfrac{3}{18} = 5\dfrac{1}{6}$

18. $6\dfrac{13}{18} - 4\dfrac{3}{18} = 2\dfrac{10}{18} = 2\dfrac{5}{9}$

175

Multiply fractions and write the answers in simplest form.

1. $2 \times \dfrac{3}{4} = \dfrac{3}{2}$ 　　　　2. $\dfrac{3}{4} \times 8 = 6$

3. $2 \times \dfrac{3}{2} = 3$ 　　　　4. $\dfrac{2}{5} \times 5 = 2$

5. $3 \times \dfrac{5}{6} = \dfrac{5}{2}$ 　　　　6. $\dfrac{3}{7} \times 7 = 3$

7. $4 \times \dfrac{2}{6} = \dfrac{4}{3}$ 　　　　8. $\dfrac{3}{8} \times 4 = \dfrac{3}{2}$

9. $5 \times \dfrac{7}{10} = \dfrac{7}{2}$ 　　　　10. $\dfrac{5}{8} \times 12 = \dfrac{15}{2}$

11. $5 \times \dfrac{3}{10} = \dfrac{3}{2}$ 　　　　12. $\dfrac{7}{10} \times 5 = \dfrac{7}{2}$

13. $4 \times \dfrac{5}{6} = \dfrac{10}{3}$ 　　　　14. $\dfrac{4}{12} \times 3 = 1$

15. $4 \times \dfrac{7}{12} = \dfrac{7}{3}$ 　　　　16. $\dfrac{5}{9} \times 3 = \dfrac{5}{3}$

17. $6 \times \dfrac{9}{12} = \dfrac{9}{2}$ 　　　　18. $\dfrac{7}{15} \times 5 = \dfrac{7}{3}$

Multiply fractions and write the answers in simplest form.

1. $\dfrac{1}{2} \times \dfrac{2}{3} = \dfrac{1}{3}$ 　　　　2. $\dfrac{1}{3} \times \dfrac{3}{4} = \dfrac{1}{4}$

3. $\dfrac{1}{3} \times \dfrac{2}{5} = \dfrac{2}{15}$ 　　　　4. $\dfrac{1}{2} \times \dfrac{1}{3} = \dfrac{1}{6}$

5. $\dfrac{1}{2} \times \dfrac{4}{5} = \dfrac{2}{5}$ 　　　　6. $\dfrac{1}{3} \times \dfrac{3}{5} = \dfrac{1}{5}$

7. $\dfrac{1}{4} \times \dfrac{4}{7} = \dfrac{1}{7}$ 　　　　8. $\dfrac{3}{7} \times \dfrac{2}{3} = \dfrac{2}{7}$

9. $\dfrac{2}{3} \times \dfrac{3}{4} = \dfrac{1}{2}$ 　　　　10. $\dfrac{2}{3} \times \dfrac{3}{6} = \dfrac{1}{3}$

11. $\dfrac{3}{5} \times \dfrac{5}{6} = \dfrac{1}{2}$ 　　　　12. $\dfrac{4}{5} \times \dfrac{5}{8} = \dfrac{1}{2}$

13. $\dfrac{7}{8} \times \dfrac{4}{7} = \dfrac{1}{2}$ 　　　　14. $\dfrac{2}{3} \times \dfrac{4}{7} = \dfrac{8}{21}$

15. $\dfrac{3}{4} \times \dfrac{3}{5} = \dfrac{9}{20}$ 　　　　16. $\dfrac{4}{9} \times \dfrac{3}{8} = \dfrac{1}{6}$

17. $\dfrac{2}{6} \times \dfrac{9}{12} = \dfrac{1}{4}$ 　　　　18. $\dfrac{3}{12} \times \dfrac{8}{9} = \dfrac{2}{9}$

Multiply fractions and write the answers in simplest form.

1. $\dfrac{1}{3} \times \dfrac{3}{6} = \dfrac{1}{6}$ 　　　　2. $\dfrac{5}{4} \times \dfrac{1}{7} = \dfrac{5}{28}$

3. $\dfrac{3}{8} \times \dfrac{3}{5} = \dfrac{9}{40}$ 　　　　4. $\dfrac{5}{2} \times \dfrac{6}{5} = 3$

5. $\dfrac{3}{5} \times \dfrac{8}{9} = \dfrac{8}{15}$ 　　　　6. $\dfrac{3}{5} \times \dfrac{5}{3} = 1$

7. $\dfrac{1}{2} \times \dfrac{6}{8} = \dfrac{3}{8}$ 　　　　8. $\dfrac{6}{7} \times \dfrac{7}{9} = \dfrac{2}{3}$

9. $\dfrac{2}{7} \times \dfrac{3}{4} = \dfrac{3}{14}$ 　　　　10. $\dfrac{2}{6} \times \dfrac{2}{4} = \dfrac{1}{6}$

11. $\dfrac{4}{9} \times \dfrac{1}{3} = \dfrac{4}{27}$ 　　　　12. $\dfrac{1}{9} \times \dfrac{3}{4} = \dfrac{1}{12}$

13. $\dfrac{2}{5} \times \dfrac{4}{6} = \dfrac{4}{15}$ 　　　　14. $\dfrac{2}{6} \times \dfrac{8}{4} = \dfrac{2}{3}$

15. $\dfrac{3}{2} \times \dfrac{4}{6} = 1$ 　　　　16. $\dfrac{3}{7} \times \dfrac{5}{3} = \dfrac{5}{7}$

17. $\dfrac{1}{7} \times \dfrac{1}{8} = \dfrac{1}{56}$ 　　　　18. $\dfrac{3}{8} \times \dfrac{6}{12} = \dfrac{3}{16}$

Multiply fractions and write the answers in simplest form.

1. $\dfrac{3}{4} \times \dfrac{2}{5} = \dfrac{3}{10}$ 　　　　2. $\dfrac{2}{3} \times \dfrac{2}{10} = \dfrac{2}{15}$

3. $\dfrac{2}{9} \times \dfrac{9}{5} = \dfrac{2}{5}$ 　　　　4. $\dfrac{6}{11} \times \dfrac{10}{12} = \dfrac{5}{11}$

5. $\dfrac{7}{4} \times \dfrac{12}{5} = \dfrac{21}{5}$ 　　　　6. $\dfrac{2}{6} \times \dfrac{7}{2} = \dfrac{7}{6}$

7. $\dfrac{11}{5} \times \dfrac{10}{7} = \dfrac{22}{7}$ 　　　　8. $\dfrac{8}{3} \times \dfrac{1}{6} = \dfrac{4}{9}$

9. $\dfrac{5}{9} \times \dfrac{12}{5} = \dfrac{4}{3}$ 　　　　10. $\dfrac{7}{3} \times \dfrac{3}{4} = \dfrac{7}{4}$

11. $\dfrac{9}{8} \times \dfrac{12}{3} = \dfrac{9}{2}$ 　　　　12. $\dfrac{9}{11} \times \dfrac{7}{2} = \dfrac{63}{22}$

13. $\dfrac{5}{7} \times \dfrac{12}{4} = \dfrac{15}{7}$ 　　　　14. $\dfrac{9}{10} \times \dfrac{3}{11} = \dfrac{27}{110}$

15. $\dfrac{8}{3} \times \dfrac{3}{8} = 1$ 　　　　16. $\dfrac{6}{12} \times \dfrac{8}{7} = \dfrac{4}{7}$

17. $\dfrac{6}{4} \times \dfrac{8}{9} = \dfrac{4}{3}$ 　　　　18. $\dfrac{3}{12} \times \dfrac{9}{6} = \dfrac{3}{8}$

Multiply fractions and write the answers in simplest form.

1. $\dfrac{4}{5} \times \dfrac{10}{6} = \dfrac{4}{3}$

2. $\dfrac{7}{12} \times \dfrac{4}{7} = \dfrac{1}{3}$

3. $\dfrac{3}{7} \times \dfrac{14}{15} = \dfrac{2}{5}$

4. $\dfrac{6}{2} \times \dfrac{4}{13} = \dfrac{12}{13}$

5. $\dfrac{12}{7} \times \dfrac{14}{7} = \dfrac{24}{7}$

6. $\dfrac{2}{6} \times \dfrac{11}{6} = \dfrac{11}{18}$

7. $\dfrac{5}{3} \times \dfrac{12}{8} = \dfrac{5}{2}$

8. $\dfrac{2}{14} \times \dfrac{10}{6} = \dfrac{5}{21}$

9. $\dfrac{7}{9} \times \dfrac{3}{14} = \dfrac{1}{6}$

10. $\dfrac{14}{8} \times \dfrac{12}{7} = 3$

11. $\dfrac{6}{5} \times \dfrac{2}{12} = \dfrac{1}{5}$

12. $\dfrac{9}{15} \times \dfrac{12}{8} = \dfrac{9}{10}$

13. $\dfrac{6}{15} \times \dfrac{10}{11} = \dfrac{4}{11}$

14. $\dfrac{13}{5} \times \dfrac{10}{8} = \dfrac{13}{4}$

15. $\dfrac{3}{11} \times \dfrac{10}{6} = \dfrac{5}{11}$

16. $\dfrac{8}{12} \times \dfrac{6}{4} = 1$

17. $\dfrac{5}{8} \times \dfrac{5}{15} = \dfrac{5}{24}$

18. $\dfrac{7}{2} \times \dfrac{8}{9} = \dfrac{28}{9}$

Multiply fractions and write the answers in simplest form.

1. $\dfrac{7}{6} \times \dfrac{6}{3} = \dfrac{7}{3}$

2. $\dfrac{14}{13} \times \dfrac{11}{7} = \dfrac{22}{13}$

3. $\dfrac{11}{8} \times \dfrac{7}{22} = \dfrac{7}{16}$

4. $\dfrac{10}{4} \times \dfrac{14}{8} = \dfrac{35}{8}$

5. $\dfrac{12}{17} \times \dfrac{7}{16} = \dfrac{21}{68}$

6. $\dfrac{2}{11} \times \dfrac{22}{7} = \dfrac{4}{7}$

7. $\dfrac{2}{15} \times \dfrac{9}{13} = \dfrac{6}{65}$

8. $\dfrac{3}{5} \times \dfrac{10}{9} = \dfrac{2}{3}$

9. $\dfrac{10}{9} \times \dfrac{9}{10} = 1$

10. $\dfrac{14}{16} \times \dfrac{4}{7} = \dfrac{1}{2}$

11. $\dfrac{4}{18} \times \dfrac{9}{10} = \dfrac{1}{5}$

12. $\dfrac{12}{5} \times \dfrac{2}{16} = \dfrac{3}{10}$

13. $\dfrac{16}{17} \times \dfrac{8}{12} = \dfrac{32}{51}$

14. $\dfrac{5}{16} \times \dfrac{6}{15} = \dfrac{1}{8}$

15. $\dfrac{3}{8} \times \dfrac{4}{9} = \dfrac{1}{6}$

16. $\dfrac{6}{2} \times \dfrac{4}{14} = \dfrac{6}{7}$

17. $\dfrac{4}{16} \times \dfrac{8}{11} = \dfrac{2}{11}$

18. $\dfrac{4}{17} \times \dfrac{34}{16} = \dfrac{1}{2}$

Multiply fractions and write the answers in simplest form.

1. $\dfrac{4}{5} \times \dfrac{6}{14} = \dfrac{12}{35}$

2. $\dfrac{12}{13} \times \dfrac{13}{24} = \dfrac{1}{2}$

3. $\dfrac{9}{10} \times \dfrac{14}{21} = \dfrac{3}{5}$

4. $\dfrac{11}{3} \times \dfrac{9}{33} = 1$

5. $\dfrac{17}{14} \times \dfrac{7}{34} = \dfrac{1}{4}$

6. $\dfrac{16}{3} \times \dfrac{15}{24} = \dfrac{10}{3}$

7. $\dfrac{13}{24} \times \dfrac{12}{7} = \dfrac{13}{14}$

8. $\dfrac{12}{21} \times \dfrac{14}{9} = \dfrac{8}{9}$

9. $\dfrac{22}{19} \times \dfrac{38}{11} = 4$

10. $\dfrac{8}{22} \times \dfrac{2}{11} = \dfrac{8}{121}$

11. $\dfrac{8}{20} \times \dfrac{13}{8} = \dfrac{13}{10}$

12. $\dfrac{3}{4} \times \dfrac{11}{9} = \dfrac{11}{12}$

13. $\dfrac{17}{9} \times \dfrac{12}{17} = \dfrac{4}{3}$

14. $\dfrac{16}{13} \times \dfrac{13}{17} = \dfrac{16}{17}$

15. $\dfrac{5}{15} \times \dfrac{9}{10} = \dfrac{3}{10}$

16. $\dfrac{11}{17} \times \dfrac{19}{11} = \dfrac{19}{17}$

17. $\dfrac{13}{16} \times \dfrac{14}{7} = \dfrac{13}{8}$

18. $\dfrac{2}{8} \times \dfrac{10}{11} = \dfrac{5}{22}$

Multiply fractions and write the answers in simplest form.

1. $\dfrac{3}{20} \times \dfrac{10}{18} = \dfrac{1}{12}$

2. $\dfrac{4}{11} \times \dfrac{33}{12} = 1$

3. $\dfrac{12}{5} \times \dfrac{15}{24} = \dfrac{3}{2}$

4. $\dfrac{18}{12} \times \dfrac{17}{6} = \dfrac{17}{4}$

5. $\dfrac{9}{30} \times \dfrac{18}{3} = \dfrac{9}{5}$

6. $\dfrac{15}{9} \times \dfrac{18}{30} = 1$

7. $\dfrac{22}{3} \times \dfrac{9}{33} = 2$

8. $\dfrac{4}{22} \times \dfrac{33}{2} = 3$

9. $\dfrac{12}{9} \times \dfrac{15}{3} = \dfrac{20}{3}$

10. $\dfrac{23}{4} \times \dfrac{2}{23} = \dfrac{1}{2}$

11. $\dfrac{12}{11} \times \dfrac{17}{24} = \dfrac{17}{22}$

12. $\dfrac{18}{28} \times \dfrac{7}{9} = \dfrac{1}{2}$

13. $\dfrac{10}{19} \times \dfrac{38}{40} = \dfrac{1}{2}$

14. $\dfrac{21}{19} \times \dfrac{19}{28} = \dfrac{3}{4}$

15. $\dfrac{10}{3} \times \dfrac{2}{4} = \dfrac{5}{3}$

16. $\dfrac{16}{29} \times \dfrac{2}{8} = \dfrac{4}{29}$

17. $\dfrac{13}{17} \times \dfrac{34}{26} = 1$

18. $\dfrac{5}{10} \times \dfrac{30}{15} = 1$

Multiply fractions and write the answers in simplest form.

1. $\dfrac{13}{24} \times \dfrac{12}{5} = \dfrac{13}{10}$

2. $\dfrac{12}{15} \times \dfrac{21}{18} = \dfrac{14}{15}$

3. $\dfrac{18}{12} \times \dfrac{9}{10} = \dfrac{27}{20}$

4. $\dfrac{16}{5} \times \dfrac{15}{12} = 4$

5. $\dfrac{9}{25} \times \dfrac{10}{3} = \dfrac{6}{5}$

6. $\dfrac{2}{12} \times \dfrac{3}{7} = \dfrac{1}{14}$

7. $\dfrac{7}{6} \times \dfrac{3}{21} = \dfrac{1}{6}$

8. $\dfrac{4}{9} \times \dfrac{27}{12} = 1$

9. $\dfrac{17}{24} \times \dfrac{8}{5} = \dfrac{17}{15}$

10. $\dfrac{5}{3} \times \dfrac{8}{10} = \dfrac{4}{3}$

11. $\dfrac{11}{18} \times \dfrac{9}{33} = \dfrac{1}{6}$

12. $\dfrac{4}{22} \times \dfrac{11}{12} = \dfrac{1}{6}$

13. $\dfrac{13}{20} \times \dfrac{10}{39} = \dfrac{1}{6}$

14. $\dfrac{5}{27} \times \dfrac{3}{20} = \dfrac{1}{36}$

15. $\dfrac{5}{10} \times \dfrac{20}{8} = \dfrac{5}{4}$

16. $\dfrac{9}{15} \times \dfrac{5}{6} = \dfrac{1}{2}$

17. $\dfrac{27}{18} \times \dfrac{9}{18} = \dfrac{3}{4}$

18. $\dfrac{12}{21} \times \dfrac{7}{18} = \dfrac{2}{9}$

Multiply fractions and write the answers in simplest form.

1. $\dfrac{22}{27} \times \dfrac{9}{10} = \dfrac{11}{15}$

2. $\dfrac{26}{4} \times \dfrac{14}{13} = 7$

3. $\dfrac{16}{25} \times \dfrac{15}{8} = \dfrac{6}{5}$

4. $\dfrac{10}{28} \times \dfrac{14}{5} = 1$

5. $\dfrac{23}{14} \times \dfrac{21}{23} = \dfrac{3}{2}$

6. $\dfrac{25}{6} \times \dfrac{24}{15} = \dfrac{20}{3}$

7. $\dfrac{7}{22} \times \dfrac{33}{14} = \dfrac{3}{4}$

8. $\dfrac{11}{28} \times \dfrac{21}{33} = \dfrac{1}{4}$

9. $\dfrac{13}{2} \times \dfrac{8}{39} = \dfrac{4}{3}$

10. $\dfrac{4}{28} \times \dfrac{14}{2} = 1$

11. $\dfrac{16}{17} \times \dfrac{34}{22} = \dfrac{16}{11}$

12. $\dfrac{13}{4} \times \dfrac{12}{39} = 1$

13. $\dfrac{14}{16} \times \dfrac{8}{7} = 1$

14. $\dfrac{26}{17} \times \dfrac{8}{39} = \dfrac{16}{51}$

15. $\dfrac{16}{17} \times \dfrac{17}{18} = \dfrac{8}{9}$

16. $\dfrac{3}{19} \times \dfrac{57}{6} = \dfrac{3}{2}$

17. $\dfrac{27}{36} \times \dfrac{12}{18} = \dfrac{1}{2}$

18. $\dfrac{8}{21} \times \dfrac{42}{24} = \dfrac{2}{3}$

Divide fractions and write the answers in simplest form.

1. $\dfrac{2}{3} \div 3 = \dfrac{2}{9}$

2. $2 \div \dfrac{2}{3} = 3$

3. $\dfrac{1}{2} \div 2 = \dfrac{1}{4}$

4. $3 \div \dfrac{1}{3} = 9$

5. $\dfrac{3}{4} \div 3 = \dfrac{1}{4}$

6. $4 \div \dfrac{8}{5} = \dfrac{5}{2}$

7. $\dfrac{2}{5} \div 4 = \dfrac{1}{10}$

8. $6 \div \dfrac{12}{7} = \dfrac{7}{2}$

9. $\dfrac{1}{6} \div 3 = \dfrac{1}{18}$

10. $5 \div \dfrac{10}{15} = \dfrac{15}{2}$

11. $\dfrac{2}{4} \div 3 = \dfrac{1}{6}$

12. $4 \div \dfrac{8}{7} = \dfrac{7}{2}$

13. $\dfrac{3}{5} \div 6 = \dfrac{1}{10}$

14. $8 \div \dfrac{12}{5} = \dfrac{10}{3}$

15. $\dfrac{4}{7} \div 6 = \dfrac{2}{21}$

16. $9 \div \dfrac{6}{4} = 6$

17. $\dfrac{4}{5} \div 12 = \dfrac{1}{15}$

18. $10 \div \dfrac{20}{3} = \dfrac{3}{2}$

Divide fractions and write the answers in simplest form.

1. $\dfrac{1}{8} \div \dfrac{3}{8} = \dfrac{1}{3}$

2. $\dfrac{3}{2} \div \dfrac{7}{2} = \dfrac{3}{7}$

3. $\dfrac{5}{7} \div \dfrac{5}{6} = \dfrac{6}{7}$

4. $\dfrac{7}{5} \div \dfrac{3}{5} = \dfrac{7}{3}$

5. $\dfrac{3}{4} \div \dfrac{2}{3} = \dfrac{9}{8}$

6. $\dfrac{2}{7} \div \dfrac{7}{6} = \dfrac{12}{49}$

7. $\dfrac{2}{6} \div \dfrac{8}{3} = \dfrac{1}{8}$

8. $\dfrac{7}{6} \div \dfrac{3}{5} = \dfrac{35}{18}$

9. $\dfrac{2}{5} \div \dfrac{10}{3} = \dfrac{3}{25}$

10. $\dfrac{6}{4} \div \dfrac{3}{5} = \dfrac{5}{2}$

11. $\dfrac{3}{8} \div \dfrac{6}{4} = \dfrac{1}{4}$

12. $\dfrac{8}{3} \div \dfrac{8}{5} = \dfrac{5}{3}$

13. $\dfrac{2}{7} \div \dfrac{6}{2} = \dfrac{2}{21}$

14. $\dfrac{3}{5} \div \dfrac{8}{6} = \dfrac{9}{20}$

15. $\dfrac{4}{3} \div \dfrac{8}{7} = \dfrac{7}{6}$

16. $\dfrac{6}{5} \div \dfrac{8}{9} = \dfrac{27}{20}$

17. $\dfrac{4}{5} \div \dfrac{4}{6} = \dfrac{6}{5}$

18. $\dfrac{5}{2} \div \dfrac{5}{3} = \dfrac{3}{2}$

Divide fractions and write the answers in simplest form.

1. $\dfrac{3}{5} \div \dfrac{4}{3} = \dfrac{9}{20}$

2. $\dfrac{4}{6} \div \dfrac{4}{8} = \dfrac{4}{3}$

3. $\dfrac{9}{12} \div \dfrac{10}{4} = \dfrac{3}{10}$

4. $\dfrac{1}{6} \div \dfrac{11}{12} = \dfrac{2}{11}$

5. $\dfrac{2}{9} \div \dfrac{12}{5} = \dfrac{5}{54}$

6. $\dfrac{10}{12} \div \dfrac{4}{7} = \dfrac{35}{24}$

7. $\dfrac{1}{2} \div \dfrac{6}{7} = \dfrac{7}{12}$

8. $\dfrac{6}{7} \div \dfrac{7}{3} = \dfrac{18}{49}$

9. $\dfrac{9}{6} \div \dfrac{3}{5} = \dfrac{5}{2}$

10. $\dfrac{10}{7} \div \dfrac{3}{6} = \dfrac{20}{7}$

11. $\dfrac{4}{10} \div \dfrac{12}{9} = \dfrac{3}{10}$

12. $\dfrac{2}{10} \div \dfrac{3}{10} = \dfrac{2}{3}$

13. $\dfrac{4}{12} \div \dfrac{10}{3} = \dfrac{1}{10}$

14. $\dfrac{3}{9} \div \dfrac{10}{12} = \dfrac{2}{5}$

15. $\dfrac{2}{10} \div \dfrac{10}{12} = \dfrac{6}{25}$

16. $\dfrac{11}{4} \div \dfrac{9}{6} = \dfrac{11}{6}$

17. $\dfrac{11}{7} \div \dfrac{8}{14} = \dfrac{11}{4}$

18. $\dfrac{3}{7} \div \dfrac{4}{10} = \dfrac{15}{14}$

Divide fractions and write the answers in simplest form.

1. $\dfrac{4}{12} \div \dfrac{4}{5} = \dfrac{5}{12}$

2. $\dfrac{5}{11} \div \dfrac{11}{4} = \dfrac{20}{121}$

3. $\dfrac{9}{8} \div \dfrac{2}{7} = \dfrac{63}{16}$

4. $\dfrac{10}{9} \div \dfrac{12}{9} = \dfrac{5}{6}$

5. $\dfrac{6}{2} \div \dfrac{7}{8} = \dfrac{24}{7}$

6. $\dfrac{11}{6} \div \dfrac{9}{10} = \dfrac{55}{27}$

7. $\dfrac{5}{8} \div \dfrac{4}{5} = \dfrac{25}{32}$

8. $\dfrac{2}{7} \div \dfrac{10}{3} = \dfrac{3}{35}$

9. $\dfrac{6}{9} \div \dfrac{11}{8} = \dfrac{16}{33}$

10. $\dfrac{2}{8} \div \dfrac{5}{4} = \dfrac{1}{5}$

11. $\dfrac{6}{5} \div \dfrac{3}{5} = 2$

12. $\dfrac{8}{4} \div \dfrac{5}{9} = \dfrac{18}{5}$

13. $\dfrac{10}{7} \div \dfrac{8}{7} = \dfrac{5}{4}$

14. $\dfrac{6}{10} \div \dfrac{7}{3} = \dfrac{9}{35}$

15. $\dfrac{5}{9} \div \dfrac{9}{4} = \dfrac{20}{81}$

16. $\dfrac{4}{10} \div \dfrac{2}{5} = 1$

17. $\dfrac{5}{11} \div \dfrac{2}{5} = \dfrac{25}{22}$

18. $\dfrac{3}{11} \div \dfrac{12}{3} = \dfrac{3}{44}$

Divide fractions and write the answers in simplest form.

1. $\dfrac{1}{7} \div \dfrac{3}{4} = \dfrac{4}{21}$

2. $\dfrac{3}{5} \div \dfrac{9}{12} = \dfrac{4}{5}$

3. $\dfrac{4}{6} \div \dfrac{6}{7} = \dfrac{7}{9}$

4. $\dfrac{5}{11} \div \dfrac{7}{11} = \dfrac{5}{7}$

5. $\dfrac{12}{11} \div \dfrac{5}{10} = \dfrac{24}{11}$

6. $\dfrac{2}{7} \div \dfrac{8}{6} = \dfrac{3}{14}$

7. $\dfrac{3}{12} \div \dfrac{2}{12} = \dfrac{3}{2}$

8. $\dfrac{3}{5} \div \dfrac{5}{7} = \dfrac{21}{25}$

9. $\dfrac{2}{6} \div \dfrac{4}{7} = \dfrac{7}{12}$

10. $\dfrac{5}{12} \div \dfrac{10}{6} = \dfrac{1}{4}$

11. $\dfrac{1}{4} \div \dfrac{8}{11} = \dfrac{11}{32}$

12. $\dfrac{3}{8} \div \dfrac{3}{4} = \dfrac{1}{2}$

13. $\dfrac{4}{10} \div \dfrac{8}{6} = \dfrac{3}{10}$

14. $\dfrac{12}{7} \div \dfrac{9}{11} = \dfrac{44}{21}$

15. $\dfrac{3}{8} \div \dfrac{6}{5} = \dfrac{5}{16}$

16. $\dfrac{2}{4} \div \dfrac{3}{6} = 1$

17. $\dfrac{1}{3} \div \dfrac{6}{8} = \dfrac{4}{9}$

18. $\dfrac{10}{7} \div \dfrac{4}{3} = \dfrac{15}{14}$

Divide fractions and write the answers in simplest form.

1. $\dfrac{13}{12} \div \dfrac{9}{13} = \dfrac{169}{108}$

2. $\dfrac{5}{8} \div \dfrac{7}{11} = \dfrac{55}{56}$

3. $\dfrac{8}{12} \div \dfrac{8}{5} = \dfrac{5}{12}$

4. $\dfrac{6}{10} \div \dfrac{3}{12} = \dfrac{12}{5}$

5. $\dfrac{14}{11} \div \dfrac{7}{5} = \dfrac{10}{11}$

6. $\dfrac{13}{11} \div \dfrac{11}{3} = \dfrac{39}{121}$

7. $\dfrac{9}{10} \div \dfrac{5}{8} = \dfrac{36}{25}$

8. $\dfrac{15}{2} \div \dfrac{6}{5} = \dfrac{25}{4}$

9. $\dfrac{1}{12} \div \dfrac{4}{14} = \dfrac{7}{24}$

10. $\dfrac{2}{9} \div \dfrac{12}{7} = \dfrac{7}{54}$

11. $\dfrac{7}{3} \div \dfrac{14}{5} = \dfrac{5}{6}$

12. $\dfrac{10}{13} \div \dfrac{8}{5} = \dfrac{25}{52}$

13. $\dfrac{14}{10} \div \dfrac{14}{10} = 1$

14. $\dfrac{12}{15} \div \dfrac{4}{10} = 2$

15. $\dfrac{15}{10} \div \dfrac{12}{5} = \dfrac{5}{8}$

16. $\dfrac{4}{3} \div \dfrac{12}{7} = \dfrac{7}{9}$

17. $\dfrac{10}{11} \div \dfrac{14}{15} = \dfrac{75}{77}$

18. $\dfrac{14}{2} \div \dfrac{7}{13} = 13$

Divide fractions and write the answers in simplest form.

1. $\dfrac{5}{6} \div \dfrac{12}{15} = \dfrac{25}{24}$

2. $\dfrac{10}{11} \div \dfrac{6}{7} = \dfrac{35}{33}$

3. $\dfrac{6}{3} \div \dfrac{3}{9} = 6$

4. $\dfrac{1}{7} \div \dfrac{6}{3} = \dfrac{1}{14}$

5. $\dfrac{7}{15} \div \dfrac{6}{5} = \dfrac{7}{18}$

6. $\dfrac{15}{10} \div \dfrac{2}{3} = \dfrac{9}{4}$

7. $\dfrac{3}{13} \div \dfrac{26}{14} = \dfrac{21}{169}$

8. $\dfrac{13}{7} \div \dfrac{39}{5} = \dfrac{5}{21}$

9. $\dfrac{11}{10} \div \dfrac{6}{14} = \dfrac{77}{30}$

10. $\dfrac{15}{14} \div \dfrac{15}{7} = \dfrac{1}{2}$

11. $\dfrac{7}{2} \div \dfrac{8}{9} = \dfrac{63}{16}$

12. $\dfrac{14}{12} \div \dfrac{7}{3} = \dfrac{1}{2}$

13. $\dfrac{14}{3} \div \dfrac{9}{10} = \dfrac{140}{27}$

14. $\dfrac{8}{12} \div \dfrac{11}{3} = \dfrac{2}{11}$

15. $\dfrac{9}{5} \div \dfrac{4}{12} = \dfrac{27}{5}$

16. $\dfrac{4}{12} \div \dfrac{5}{14} = \dfrac{14}{15}$

17. $\dfrac{12}{11} \div \dfrac{9}{6} = \dfrac{8}{11}$

18. $\dfrac{7}{8} \div \dfrac{7}{9} = \dfrac{9}{8}$

Divide fractions and write the answers in simplest form.

1. $\dfrac{9}{10} \div \dfrac{6}{8} = \dfrac{6}{5}$

2. $\dfrac{13}{14} \div \dfrac{26}{28} = 1$

3. $\dfrac{10}{16} \div \dfrac{12}{7} = \dfrac{35}{96}$

4. $\dfrac{9}{20} \div \dfrac{12}{30} = \dfrac{9}{8}$

5. $\dfrac{5}{14} \div \dfrac{8}{7} = \dfrac{5}{16}$

6. $\dfrac{8}{9} \div \dfrac{24}{18} = \dfrac{2}{3}$

7. $\dfrac{11}{6} \div \dfrac{10}{22} = \dfrac{121}{30}$

8. $\dfrac{9}{17} \div \dfrac{36}{34} = \dfrac{1}{2}$

9. $\dfrac{14}{3} \div \dfrac{28}{5} = \dfrac{5}{6}$

10. $\dfrac{7}{6} \div \dfrac{8}{6} = \dfrac{7}{8}$

11. $\dfrac{2}{11} \div \dfrac{3}{22} = \dfrac{4}{3}$

12. $\dfrac{14}{6} \div \dfrac{18}{5} = \dfrac{35}{54}$

13. $\dfrac{8}{5} \div \dfrac{12}{15} = 2$

14. $\dfrac{7}{17} \div \dfrac{5}{51} = \dfrac{21}{5}$

15. $\dfrac{3}{13} \div \dfrac{9}{39} = 1$

16. $\dfrac{6}{7} \div \dfrac{12}{7} = \dfrac{1}{2}$

17. $\dfrac{12}{10} \div \dfrac{12}{5} = \dfrac{1}{2}$

18. $\dfrac{16}{19} \div \dfrac{8}{19} = 2$

Divide fractions and write the answers in simplest form.

1. $\dfrac{18}{9} \div \dfrac{2}{13} = 13$

2. $\dfrac{18}{5} \div \dfrac{6}{35} = 21$

3. $\dfrac{17}{7} \div \dfrac{17}{11} = \dfrac{11}{7}$

4. $\dfrac{11}{16} \div \dfrac{33}{24} = \dfrac{1}{2}$

5. $\dfrac{4}{14} \div \dfrac{18}{7} = \dfrac{1}{9}$

6. $\dfrac{6}{9} \div \dfrac{4}{3} = \dfrac{1}{2}$

7. $\dfrac{11}{18} \div \dfrac{22}{12} = \dfrac{1}{3}$

8. $\dfrac{14}{3} \div \dfrac{8}{5} = \dfrac{35}{12}$

9. $\dfrac{4}{15} \div \dfrac{18}{7} = \dfrac{14}{135}$

10. $\dfrac{4}{19} \div \dfrac{12}{38} = \dfrac{2}{3}$

11. $\dfrac{8}{13} \div \dfrac{6}{26} = \dfrac{8}{3}$

12. $\dfrac{16}{14} \div \dfrac{4}{8} = \dfrac{16}{7}$

13. $\dfrac{14}{18} \div \dfrac{7}{6} = \dfrac{2}{3}$

14. $\dfrac{7}{4} \div \dfrac{4}{6} = \dfrac{21}{8}$

15. $\dfrac{20}{6} \div \dfrac{5}{12} = 8$

16. $\dfrac{18}{11} \div \dfrac{2}{22} = 18$

17. $\dfrac{16}{17} \div \dfrac{4}{34} = 8$

18. $\dfrac{13}{8} \div \dfrac{13}{2} = \dfrac{1}{4}$

Divide fractions and write the answers in simplest form.

1. $\dfrac{6}{18} \div \dfrac{12}{3} = \dfrac{1}{12}$

2. $\dfrac{8}{20} \div \dfrac{7}{15} = \dfrac{6}{7}$

3. $\dfrac{16}{10} \div \dfrac{8}{5} = 1$

4. $\dfrac{5}{8} \div \dfrac{7}{8} = \dfrac{5}{7}$

5. $\dfrac{7}{4} \div \dfrac{8}{12} = \dfrac{21}{8}$

6. $\dfrac{7}{5} \div \dfrac{6}{15} = \dfrac{7}{2}$

7. $\dfrac{3}{13} \div \dfrac{9}{5} = \dfrac{5}{39}$

8. $\dfrac{16}{15} \div \dfrac{11}{15} = \dfrac{16}{11}$

9. $\dfrac{7}{11} \div \dfrac{14}{4} = \dfrac{2}{11}$

10. $\dfrac{15}{17} \div \dfrac{5}{6} = \dfrac{18}{17}$

11. $\dfrac{4}{10} \div \dfrac{2}{11} = \dfrac{11}{5}$

12. $\dfrac{3}{2} \div \dfrac{4}{7} = \dfrac{21}{8}$

13. $\dfrac{9}{5} \div \dfrac{11}{10} = \dfrac{18}{11}$

14. $\dfrac{19}{9} \div \dfrac{38}{6} = \dfrac{1}{3}$

15. $\dfrac{4}{12} \div \dfrac{8}{17} = \dfrac{17}{24}$

16. $\dfrac{13}{8} \div \dfrac{39}{12} = \dfrac{1}{2}$

17. $\dfrac{2}{10} \div \dfrac{11}{9} = \dfrac{9}{55}$

18. $\dfrac{5}{15} \div \dfrac{15}{30} = \dfrac{2}{3}$

Change the fractions to decimals and round to the nearest thousandth if necessary.

1. $\dfrac{1}{4} = 0.25$ 2. $\dfrac{1}{5} = 0.2$ 3. $\dfrac{3}{5} = 0.6$

4. $\dfrac{2}{5} = 0.4$ 5. $\dfrac{2}{4} = 0.5$ 6. $\dfrac{6}{8} = 0.75$

7. $\dfrac{3}{6} = 0.5$ 8. $\dfrac{4}{8} = 0.5$ 9. $\dfrac{3}{4} = 0.75$

10. $\dfrac{2}{8} = 0.25$ 11. $\dfrac{1}{10} = 0.1$ 12. $\dfrac{5}{8} = 0.625$

13. $\dfrac{2}{10} = 0.2$ 14. $\dfrac{3}{8} = 0.375$ 15. $\dfrac{4}{5} = 0.8$

16. $\dfrac{6}{10} = 0.6$ 17. $\dfrac{3}{10} = 0.3$ 18. $\dfrac{7}{8} = 0.875$

19. $\dfrac{3}{12} = 0.25$ 20. $\dfrac{5}{10} = 0.5$ 21. $\dfrac{4}{10} = 0.4$

22. $\dfrac{4}{20} = 0.2$ 23. $\dfrac{6}{12} = 0.5$ 24. $\dfrac{9}{12} = 0.75$

25. $\dfrac{6}{16} = 0.375$ 26. $\dfrac{6}{15} = 0.4$ 27. $\dfrac{4}{16} = 0.25$

Change the fractions to decimals and round to the nearest thousandth if necessary.

1. $\dfrac{1}{4} = 0.25$ 2. $\dfrac{1}{2} = 0.5$ 3. $\dfrac{1}{5} = 0.2$

4. $\dfrac{3}{4} = 0.75$ 5. $\dfrac{2}{5} = 0.4$ 6. $\dfrac{2}{4} = 0.5$

7. $\dfrac{3}{5} = 0.6$ 8. $\dfrac{2}{8} = 0.25$ 9. $\dfrac{4}{5} = 0.8$

10. $\dfrac{1}{8} = 0.125$ 11. $\dfrac{4}{20} = 0.2$ 12. $\dfrac{3}{8} = 0.375$

13. $\dfrac{4}{8} = 0.5$ 14. $\dfrac{16}{20} = 0.8$ 15. $\dfrac{6}{8} = 0.75$

16. $\dfrac{7}{8} = 0.875$ 17. $\dfrac{5}{8} = 0.625$ 18. $\dfrac{14}{20} = 0.7$

19. $\dfrac{2}{20} = 0.1$ 20. $\dfrac{9}{12} = 0.75$ 21. $\dfrac{5}{20} = 0.25$

22. $\dfrac{3}{20} = 0.15$ 23. $\dfrac{3}{25} = 0.12$ 24. $\dfrac{1}{20} = 0.05$

25. $\dfrac{4}{25} = 0.16$ 26. $\dfrac{6}{25} = 0.24$ 27. $\dfrac{5}{25} = 0.2$

Change the fractions to decimals and round to the nearest thousandth if necessary.

1. $\dfrac{3}{2} = 1.5$ 2. $\dfrac{3}{4} = 0.75$ 3. $\dfrac{5}{4} = 1.25$

4. $\dfrac{6}{5} = 1.2$ 5. $\dfrac{2}{5} = 0.4$ 6. $\dfrac{4}{5} = 0.8$

7. $\dfrac{9}{4} = 2.25$ 8. $\dfrac{7}{5} = 1.4$ 9. $\dfrac{5}{2} = 2.5$

10. $\dfrac{8}{5} = 1.6$ 11. $\dfrac{6}{4} = 1.5$ 12. $\dfrac{8}{4} = 2$

13. $\dfrac{3}{8} = 0.375$ 14. $\dfrac{7}{4} = 1.75$ 15. $\dfrac{12}{8} = 1.5$

16. $\dfrac{9}{6} = 1.5$ 17. $\dfrac{10}{8} = 1.25$ 18. $\dfrac{9}{5} = 1.8$

19. $\dfrac{12}{8} = 1.5$ 20. $\dfrac{12}{10} = 1.2$ 21. $\dfrac{9}{8} = 1.125$

22. $\dfrac{14}{10} = 1.4$ 23. $\dfrac{13}{5} = 2.6$ 24. $\dfrac{14}{10} = 1.4$

25. $\dfrac{11}{5} = 2.2$ 26. $\dfrac{13}{8} = 1.625$ 27. $\dfrac{13}{5} = 2.6$

Change the fractions to decimals and round to the nearest thousandth if necessary.

1. $\dfrac{7}{5} = 1.4$ 2. $\dfrac{3}{5} = 0.6$ 3. $\dfrac{3}{4} = 0.75$

4. $\dfrac{1}{4} = 0.25$ 5. $\dfrac{2}{4} = 0.5$ 6. $\dfrac{9}{5} = 1.8$

7. $\dfrac{3}{20} = 0.15$ 8. $\dfrac{6}{5} = 1.2$ 9. $\dfrac{9}{6} = 1.5$

10. $\dfrac{7}{4} = 1.75$ 11. $\dfrac{10}{4} = 2.5$ 12. $\dfrac{13}{4} = 3.25$

13. $\dfrac{4}{25} = 0.16$ 14. $\dfrac{14}{5} = 2.8$ 15. $\dfrac{1}{20} = 0.05$

16. $\dfrac{3}{10} = 0.3$ 17. $\dfrac{11}{10} = 1.1$ 18. $\dfrac{12}{5} = 2.4$

19. $\dfrac{9}{4} = 2.25$ 20. $\dfrac{11}{4} = 2.75$ 21. $\dfrac{3}{20} = 0.15$

22. $\dfrac{5}{25} = 0.2$ 23. $\dfrac{6}{25} = 0.24$ 24. $\dfrac{11}{20} = 0.55$

25. $\dfrac{3}{50} = 0.06$ 26. $\dfrac{6}{50} = 0.12$ 27. $\dfrac{7}{25} = 0.28$

Change the fractions to decimals and round to the nearest thousandth if necessary.

1. $\frac{6}{5} = 1.2$ 2. $\frac{6}{4} = 1.5$ 3. $\frac{7}{2} = 3.5$

4. $\frac{9}{4} = 2.25$ 5. $\frac{8}{5} = 1.6$ 6. $\frac{7}{4} = 1.75$

7. $\frac{9}{2} = 4.5$ 8. $\frac{10}{4} = 2.5$ 9. $\frac{13}{5} = 2.6$

10. $\frac{16}{10} = 1.6$ 11. $\frac{17}{2} = 8.5$ 12. $\frac{11}{4} = 2.75$

13. $\frac{13}{2} = 6.5$ 14. $\frac{15}{4} = 3.75$ 15. $\frac{9}{25} = 0.36$

16. $\frac{15}{6} = 2.5$ 17. $\frac{12}{5} = 2.4$ 18. $\frac{13}{4} = 3.25$

19. $\frac{6}{20} = 0.3$ 20. $\frac{3}{20} = 0.15$ 21. $\frac{16}{5} = 3.2$

22. $\frac{6}{25} = 0.24$ 23. $\frac{9}{8} = 1.125$ 24. $\frac{18}{12} = 1.5$

25. $\frac{17}{8} = 2.125$ 26. $\frac{7}{25} = 0.28$ 27. $\frac{15}{8} = 1.875$

Change the fractions to decimals and round to the nearest thousandth if necessary.

1. $\frac{9}{4} = 2.25$ 2. $\frac{11}{4} = 2.75$ 3. $\frac{13}{4} = 3.25$

4. $\frac{7}{2} = 3.5$ 5. $\frac{12}{8} = 1.5$ 6. $\frac{24}{16} = 1.5$

7. $\frac{11}{5} = 2.2$ 8. $\frac{11}{22} = 0.5$ 9. $\frac{17}{5} = 3.4$

10. $\frac{7}{14} = 0.5$ 11. $\frac{14}{5} = 2.8$ 12. $\frac{11}{44} = 0.25$

13. $\frac{13}{52} = 0.25$ 14. $\frac{10}{8} = 1.25$ 15. $\frac{1}{20} = 0.05$

16. $\frac{13}{25} = 0.52$ 17. $\frac{7}{20} = 0.35$ 18. $\frac{19}{5} = 3.8$

19. $\frac{11}{8} = 1.375$ 20. $\frac{15}{25} = 0.6$ 21. $\frac{14}{25} = 0.56$

22. $\frac{2}{25} = 0.08$ 23. $\frac{13}{20} = 0.65$ 24. $\frac{9}{20} = 0.45$

25. $\frac{11}{25} = 0.44$ 26. $\frac{17}{25} = 0.68$ 27. $\frac{19}{8} = 2.375$

Change the fractions to decimals and round to the nearest thousandth if necessary.

1. $\frac{13}{4} = 3.25$ 2. $\frac{4}{5} = 0.8$ 3. $\frac{17}{10} = 1.7$

4. $\frac{16}{5} = 3.2$ 5. $\frac{15}{4} = 3.75$ 6. $\frac{14}{20} = 0.7$

7. $\frac{17}{8} = 2.125$ 8. $\frac{27}{6} = 4.5$ 9. $\frac{7}{100} = 0.07$

10. $\frac{19}{20} = 0.95$ 11. $\frac{17}{5} = 3.4$ 12. $\frac{17}{4} = 4.25$

13. $\frac{13}{10} = 1.3$ 14. $\frac{29}{8} = 3.625$ 15. $\frac{9}{24} = 0.375$

16. $\frac{4}{32} = 0.125$ 17. $\frac{11}{25} = 0.44$ 18. $\frac{18}{5} = 3.6$

19. $\frac{17}{20} = 0.85$ 20. $\frac{7}{20} = 0.35$ 21. $\frac{7}{50} = 0.14$

22. $\frac{21}{8} = 2.625$ 23. $\frac{18}{20} = 0.9$ 24. $\frac{21}{5} = 4.2$

25. $\frac{17}{25} = 0.68$ 26. $\frac{1}{100} = 0.01$ 27. $\frac{4}{100} = 0.04$

Change the fractions to decimals and round to the nearest thousandth if necessary.

1. $\frac{1}{20} = 0.05$ 2. $\frac{11}{2} = 5.5$ 3. $\frac{13}{4} = 3.25$

4. $\frac{15}{2} = 7.5$ 5. $\frac{5}{20} = 0.25$ 6. $\frac{21}{6} = 3.5$

7. $\frac{29}{8} = 3.625$ 8. $\frac{5}{25} = 0.2$ 9. $\frac{11}{25} = 0.44$

10. $\frac{22}{100} = 0.22$ 11. $\frac{17}{100} = 0.17$ 12. $\frac{7}{20} = 0.35$

13. $\frac{2}{25} = 0.08$ 14. $\frac{25}{8} = 3.125$ 15. $\frac{11}{100} = 0.11$

16. $\frac{37}{100} = 0.37$ 17. $\frac{9}{25} = 0.36$ 18. $\frac{13}{25} = 0.52$

19. $\frac{21}{2} = 10.5$ 20. $\frac{23}{5} = 4.6$ 21. $\frac{9}{20} = 0.45$

22. $\frac{37}{5} = 7.4$ 23. $\frac{31}{8} = 3.875$ 24. $\frac{7}{100} = 0.07$

25. $\frac{18}{25} = 0.72$ 26. $\frac{33}{6} = 5.5$ 27. $\frac{13}{20} = 0.65$

Change the fractions to decimals and round to the nearest thousandth if necessary.

1. $\dfrac{2}{5} = 0.4$ 2. $\dfrac{1}{8} = 0.125$ 3. $\dfrac{1}{25} = 0.04$

4. $\dfrac{11}{4} = 2.75$ 5. $\dfrac{12}{5} = 2.4$ 6. $\dfrac{1}{20} = 0.05$

7. $\dfrac{11}{8} = 1.375$ 8. $\dfrac{15}{6} = 2.5$ 9. $\dfrac{19}{5} = 3.8$

10. $\dfrac{24}{16} = 1.5$ 11. $\dfrac{13}{4} = 3.25$ 12. $\dfrac{18}{24} = 0.75$

13. $\dfrac{17}{2} = 8.5$ 14. $\dfrac{9}{36} = 0.25$ 15. $\dfrac{17}{4} = 4.25$

16. $\dfrac{29}{4} = 7.25$ 17. $\dfrac{11}{100} = 0.11$ 18. $\dfrac{7}{20} = 0.35$

19. $\dfrac{8}{40} = 0.2$ 20. $\dfrac{8}{20} = 0.4$ 21. $\dfrac{3}{25} = 0.12$

22. $\dfrac{19}{25} = 0.76$ 23. $\dfrac{47}{8} = 5.875$ 24. $\dfrac{13}{52} = 0.25$

25. $\dfrac{42}{50} = 0.84$ 26. $\dfrac{33}{50} = 0.66$ 27. $\dfrac{49}{50} = 0.98$

Change the fractions to decimals and round to the nearest thousandth if necessary.

1. $\dfrac{1}{3} = 0.333$ 2. $\dfrac{14}{5} = 2.8$ 3. $\dfrac{1}{6} = 0.167$

4. $\dfrac{4}{5} = 0.8$ 5. $\dfrac{2}{3} = 0.667$ 6. $\dfrac{6}{25} = 0.24$

7. $\dfrac{19}{4} = 4.75$ 8. $\dfrac{25}{4} = 6.25$ 9. $\dfrac{2}{20} = 0.1$

10. $\dfrac{42}{5} = 8.4$ 11. $\dfrac{3}{20} = 0.15$ 12. $\dfrac{57}{8} = 7.125$

13. $\dfrac{9}{100} = 0.09$ 14. $\dfrac{43}{8} = 5.375$ 15. $\dfrac{27}{18} = 1.5$

16. $\dfrac{4}{3} = 1.333$ 17. $\dfrac{23}{2} = 11.5$ 18. $\dfrac{7}{6} = 1.167$

19. $\dfrac{22}{25} = 0.88$ 20. $\dfrac{4}{6} = 0.667$ 21. $\dfrac{65}{8} = 8.125$

22. $\dfrac{17}{20} = 0.85$ 23. $\dfrac{29}{50} = 0.58$ 24. $\dfrac{10}{3} = 3.333$

25. $\dfrac{45}{100} = 0.45$ 26. $\dfrac{23}{25} = 0.92$ 27. $\dfrac{18}{27} = 0.667$

Change decimals to fractions and write the answers in simplest form.

1. $0.2 = \dfrac{2}{10} = \dfrac{1}{5}$ 2. $0.25 = \dfrac{25}{100} = \dfrac{1}{4}$

3. $0.75 = \dfrac{75}{100} = \dfrac{3}{4}$ 4. $0.5 = \dfrac{5}{10} = \dfrac{1}{2}$

5. $0.6 = \dfrac{6}{10} = \dfrac{3}{5}$ 6. $0.1 = \dfrac{1}{10}$

7. $0.01 = \dfrac{1}{100}$ 8. $0.05 = \dfrac{5}{100} = \dfrac{1}{20}$

9. $0.4 = \dfrac{4}{10} = \dfrac{2}{5}$ 10. $0.6 = \dfrac{6}{10} = \dfrac{3}{5}$

11. $0.9 = \dfrac{9}{10}$ 12. $0.8 = \dfrac{8}{10} = \dfrac{4}{5}$

13. $0.15 = \dfrac{15}{100} = \dfrac{3}{20}$ 14. $0.35 = \dfrac{35}{100} = \dfrac{7}{20}$

15. $0.02 = \dfrac{2}{100} = \dfrac{1}{50}$ 16. $0.04 = \dfrac{4}{100} = \dfrac{1}{25}$

17. $0.65 = \dfrac{65}{100} = \dfrac{13}{20}$ 18. $0.7 = \dfrac{7}{10}$

19. $0.3 = \dfrac{3}{10}$ 20. $0.85 = \dfrac{85}{100} = \dfrac{17}{20}$

Change decimals to fractions and write the answers in simplest form.

1. $0.1 = \dfrac{1}{10}$ 2. $0.3 = \dfrac{3}{10}$

3. $0.2 = \dfrac{2}{10} = \dfrac{1}{5}$ 4. $0.4 = \dfrac{4}{10} = \dfrac{2}{5}$

5. $0.35 = \dfrac{35}{100} = \dfrac{7}{20}$ 6. $0.15 = \dfrac{15}{100} = \dfrac{3}{20}$

7. $0.25 = \dfrac{25}{100} = \dfrac{1}{4}$ 8. $0.09 = \dfrac{9}{100}$

9. $0.03 = \dfrac{3}{100}$ 10. $0.55 = \dfrac{55}{100} = \dfrac{11}{20}$

11. $0.45 = \dfrac{45}{100} = \dfrac{9}{20}$ 12. $0.11 = \dfrac{11}{100}$

13. $0.21 = \dfrac{21}{100}$ 14. $0.33 = \dfrac{33}{100}$

15. $0.65 = \dfrac{65}{100} = \dfrac{13}{20}$ 16. $0.05 = \dfrac{5}{100} = \dfrac{1}{20}$

17. $0.75 = \dfrac{75}{100} = \dfrac{3}{4}$ 18. $0.95 = \dfrac{95}{100} = \dfrac{19}{20}$

19. $0.53 = \dfrac{53}{100}$ 20. $0.07 = \dfrac{7}{100}$

Change decimals to fractions and write the answers in simplest form.

1. $0.01 = \dfrac{1}{100}$

2. $0.05 = \dfrac{5}{100} = \dfrac{1}{20}$

3. $0.8 = \dfrac{8}{10} = \dfrac{4}{5}$

4. $0.25 = \dfrac{25}{100} = \dfrac{1}{4}$

5. $0.75 = \dfrac{75}{100} = \dfrac{3}{4}$

6. $0.16 = \dfrac{16}{100} = \dfrac{4}{25}$

7. $0.125 = \dfrac{125}{1000} = \dfrac{25}{200} = \dfrac{1}{8}$

8. $0.09 = \dfrac{9}{100}$

9. $0.88 = \dfrac{88}{100} = \dfrac{22}{25}$

10. $0.08 = \dfrac{8}{100} = \dfrac{2}{25}$

11. $0.87 = \dfrac{87}{100}$

12. $0.99 = \dfrac{99}{100}$

13. $0.625 = \dfrac{625}{1000} = \dfrac{125}{200} = \dfrac{5}{8}$

14. $0.875 = \dfrac{875}{1000} = \dfrac{175}{200} = \dfrac{7}{8}$

15. $0.24 = \dfrac{24}{100} = \dfrac{6}{25}$

16. $0.15 = \dfrac{15}{100} = \dfrac{3}{20}$

17. $0.4 = \dfrac{4}{10} = \dfrac{2}{5}$

18. $0.375 = \dfrac{375}{1000} = \dfrac{75}{200} = \dfrac{3}{8}$

19. $0.57 = \dfrac{57}{100}$

20. $0.24 = \dfrac{24}{100} = \dfrac{6}{25}$

Change decimals to fractions and write the answers in simplest form.

1. $0.05 = \dfrac{5}{100} = \dfrac{1}{20}$

2. $0.3 = \dfrac{3}{10}$

3. $1.2 = \dfrac{12}{10} = \dfrac{6}{5}$

4. $1.1 = \dfrac{11}{10}$

5. $0.95 = \dfrac{95}{100} = \dfrac{19}{20}$

6. $0.875 = \dfrac{875}{1000} = \dfrac{175}{200} = \dfrac{7}{8}$

7. $2.75 = \dfrac{275}{100} = \dfrac{55}{20} = \dfrac{11}{4}$

8. $0.45 = \dfrac{45}{100} = \dfrac{9}{20}$

9. $0.85 = \dfrac{85}{100} = \dfrac{17}{20}$

10. $0.125 = \dfrac{125}{1000} = \dfrac{25}{200} = \dfrac{1}{8}$

11. $1.11 = \dfrac{111}{100}$

12. $1.6 = \dfrac{16}{10} = \dfrac{8}{5}$

13. $0.65 = \dfrac{65}{100} = \dfrac{13}{20}$

14. $0.35 = \dfrac{35}{100} = \dfrac{7}{20}$

15. $0.375 = \dfrac{375}{1000} = \dfrac{75}{200} = \dfrac{3}{8}$

16. $3.25 = \dfrac{325}{100} = \dfrac{65}{20} = \dfrac{13}{4}$

17. $0.55 = \dfrac{55}{100} = \dfrac{11}{20}$

18. $1.45 = \dfrac{145}{100} = \dfrac{29}{20}$

19. $3.65 = \dfrac{365}{100} = \dfrac{73}{20}$

20. $1.125 = \dfrac{1125}{1000} = \dfrac{225}{200} = \dfrac{9}{8}$

Change decimals to fractions and write the answers in simplest form.

1. $1.5 = \dfrac{15}{10} = \dfrac{3}{2}$

2. $2.4 = \dfrac{24}{10} = \dfrac{12}{5}$

3. $0.38 = \dfrac{38}{100} = \dfrac{19}{50}$

4. $0.44 = \dfrac{44}{100} = \dfrac{11}{25}$

5. $1.75 = \dfrac{175}{100} = \dfrac{35}{20} = \dfrac{7}{4}$

6. $4.25 = \dfrac{425}{100} = \dfrac{85}{20} = \dfrac{17}{4}$

7. $1.45 = \dfrac{145}{100} = \dfrac{29}{20}$

8. $0.05 = \dfrac{5}{100} = \dfrac{1}{20}$

9. $1.375 = \dfrac{1375}{1000} = \dfrac{275}{200} = \dfrac{11}{8}$

10. $0.68 = \dfrac{68}{100} = \dfrac{17}{25}$

11. $0.52 = \dfrac{52}{100} = \dfrac{13}{25}$

12. $2.625 = \dfrac{2625}{1000} = \dfrac{525}{200} = \dfrac{21}{8}$

13. $2.7 = \dfrac{27}{10}$

14. $3.55 = \dfrac{355}{100} = \dfrac{71}{20}$

15. $1.98 = \dfrac{198}{100} = \dfrac{99}{50}$

16. $0.87 = \dfrac{87}{100}$

17. $3.1 = \dfrac{31}{10}$

18. $1.23 = \dfrac{123}{100}$

19. $5.75 = \dfrac{575}{100} = \dfrac{115}{20} = \dfrac{23}{10}$

20. $4.99 = \dfrac{499}{100}$

Change decimals to fractions and write the answers in simplest form.

1. $1.12 = \dfrac{112}{100} = \dfrac{28}{25}$

2. $2.8 = \dfrac{28}{10} = \dfrac{14}{5}$

3. $4.125 = \dfrac{4125}{1000} = \dfrac{825}{200} = \dfrac{33}{8}$

4. $6.25 = \dfrac{625}{100} = \dfrac{25}{4}$

5. $1.96 = \dfrac{196}{100} = \dfrac{49}{25}$

6. $1.86 = \dfrac{186}{100} = \dfrac{93}{50}$

7. $0.76 = \dfrac{76}{100} = \dfrac{19}{25}$

8. $0.28 = \dfrac{28}{100} = \dfrac{7}{25}$

9. $1.58 = \dfrac{158}{100} = \dfrac{79}{50}$

10. $4.55 = \dfrac{455}{100} = \dfrac{91}{20}$

11. $5.75 = \dfrac{575}{100} = \dfrac{23}{4}$

12. $3.875 = \dfrac{3875}{1000} = \dfrac{775}{200} = \dfrac{31}{8}$

13. $2.625 = \dfrac{2625}{1000} = \dfrac{525}{200} = \dfrac{21}{8}$

14. $1.92 = \dfrac{192}{100} = \dfrac{48}{25}$

15. $1.35 = \dfrac{135}{100} = \dfrac{27}{20}$

16. $0.64 = \dfrac{64}{100} = \dfrac{16}{25}$

17. $0.78 = \dfrac{78}{100} = \dfrac{39}{50}$

18. $3.86 = \dfrac{386}{100} = \dfrac{193}{50}$

19. $2.18 = \dfrac{218}{100} = \dfrac{109}{50}$

20. $2.48 = \dfrac{248}{100} = \dfrac{62}{25}$

Lesson 8-7   Changing decimals to fractions

Change decimals to fractions and write the answers in simplest form.

1. $2.2 = \dfrac{22}{10} = \dfrac{11}{5}$

2. $0.15 = \dfrac{15}{100} = \dfrac{3}{20}$

3. $1.08 = \dfrac{108}{100} = \dfrac{27}{25}$

4. $0.4 = \dfrac{4}{10} = \dfrac{2}{5}$

5. $0.62 = \dfrac{62}{100} = \dfrac{31}{50}$

6. $3.25 = \dfrac{325}{100} = \dfrac{65}{20} = \dfrac{13}{4}$

7. $5.75 = \dfrac{575}{100} = \dfrac{115}{20} = \dfrac{23}{4}$

8. $2.6 = \dfrac{26}{10} = \dfrac{13}{5}$

9. $4.35 = \dfrac{435}{100} = \dfrac{87}{20}$

10. $0.85 = \dfrac{85}{100} = \dfrac{17}{20}$

11. $0.03 = \dfrac{3}{100}$

12. $3.04 = \dfrac{304}{100} = \dfrac{76}{25}$

13. $5.4 = \dfrac{54}{10} = \dfrac{27}{5}$

14. $4.82 = \dfrac{482}{100} = \dfrac{241}{50}$

15. $0.48 = \dfrac{48}{100} = \dfrac{12}{25}$

16. $6.75 = \dfrac{675}{100} = \dfrac{27}{4}$

17. $2.56 = \dfrac{256}{100} = \dfrac{64}{25}$

18. $1.875 = \dfrac{1875}{1000} = \dfrac{375}{200} = \dfrac{15}{8}$

19. $2.01 = \dfrac{201}{100}$

20. $5.52 = \dfrac{552}{100} = \dfrac{138}{25}$

Lesson 8-8   Changing decimals to fractions

Change decimals to fractions and write the answers in simplest form.

1. $4.09 = \dfrac{409}{100} =$

2. $0.75 = \dfrac{75}{100} = \dfrac{3}{4}$

3. $0.45 = \dfrac{45}{100} = \dfrac{9}{20}$

4. $1.53 = \dfrac{153}{100}$

5. $1.76 = \dfrac{176}{100} = \dfrac{44}{25}$

6. $3.74 = \dfrac{374}{100} = \dfrac{187}{50}$

7. $6.18 = \dfrac{618}{100} = \dfrac{309}{50}$

8. $6.3 = \dfrac{63}{10}$

9. $0.81 = \dfrac{81}{100}$

10. $4.625 = \dfrac{4625}{1000} = \dfrac{925}{200} = \dfrac{37}{8}$

11. $4.55 = \dfrac{455}{100} = \dfrac{91}{20}$

12. $8.01 = \dfrac{801}{100}$

13. $8.12 = \dfrac{812}{100} = \dfrac{203}{25}$

14. $8.02 = \dfrac{802}{100} = \dfrac{401}{50}$

15. $7.2 = \dfrac{72}{10} = \dfrac{36}{5}$

16. $9.41 = \dfrac{941}{100}$

17. $4.125 = \dfrac{4125}{1000} = \dfrac{825}{200} = \dfrac{33}{8}$

18. $1.48 = \dfrac{148}{100} = \dfrac{37}{25}$

19. $1.09 = \dfrac{109}{100}$

20. $3.98 = \dfrac{398}{100} = \dfrac{199}{50}$

Lesson 8-9   Changing decimals to fractions

Change decimals to fractions and write the answers in simplest form.

1. $0.35 = \dfrac{35}{100} = \dfrac{7}{20}$

2. $0.14 = \dfrac{14}{100} = \dfrac{7}{50}$

3. $1.68 = \dfrac{168}{100} = \dfrac{42}{25}$

4. $2.06 = \dfrac{206}{100} = \dfrac{103}{50}$

5. $0.97 = \dfrac{97}{100}$

6. $5.9 = \dfrac{59}{10}$

7. $4.82 = \dfrac{482}{100} = \dfrac{241}{50}$

8. $6.89 = \dfrac{689}{100}$

9. $8.25 = \dfrac{825}{100} = \dfrac{33}{4}$

10. $7.75 = \dfrac{775}{100} = \dfrac{31}{4}$

11. $6.92 = \dfrac{692}{100} = \dfrac{173}{25}$

12. $1.6 = \dfrac{16}{10} = \dfrac{8}{5}$

13. $7.625 = \dfrac{7625}{1000} = \dfrac{1525}{200} = \dfrac{61}{8}$

14. $2.44 = \dfrac{244}{100} = \dfrac{61}{25}$

15. $4.84 = \dfrac{484}{100} = \dfrac{121}{25}$

16. $4.16 = \dfrac{416}{100} = \dfrac{104}{25}$

17. $5.65 = \dfrac{565}{100} = \dfrac{113}{20}$

18. $8.375 = \dfrac{8375}{1000} = \dfrac{1675}{200} = \dfrac{67}{8}$

19. $7.62 = \dfrac{762}{100} = \dfrac{381}{50}$

20. $9.44 = \dfrac{944}{100} = \dfrac{236}{25}$

Lesson 8-10   Changing decimals to fractions

Change decimals to fractions and write the answers in simplest form.

1. $8.12 = \dfrac{13}{10}$

2. $2.07 = \dfrac{207}{100}$

3. $9.37 = \dfrac{937}{100}$

4. $4.2 = \dfrac{42}{10} = \dfrac{21}{5}$

5. $6.56 = \dfrac{656}{100} = \dfrac{164}{25}$

6. $5.06 = \dfrac{506}{100} = \dfrac{253}{50}$

7. $0.72 = \dfrac{72}{100} = \dfrac{18}{25}$

8. $3.7 = \dfrac{37}{10}$

9. $7.22 = \dfrac{722}{100} = \dfrac{361}{50}$

10. $0.84 = \dfrac{84}{100} = \dfrac{21}{25}$

11. $2.855 = \dfrac{2855}{1000} = \dfrac{571}{200}$

12. $8.11 = \dfrac{811}{100}$

13. $8.125 = \dfrac{8125}{1000} = \dfrac{1625}{200} = \dfrac{65}{8}$

14. $1.9 = \dfrac{19}{10}$

15. $3.62 = \dfrac{362}{100} = \dfrac{181}{50}$

16. $6.74 = \dfrac{674}{100} = \dfrac{337}{50}$

17. $5.64 = \dfrac{564}{100} = \dfrac{141}{25}$

18. $9.875 = \dfrac{9875}{1000} = \dfrac{1975}{200} = \dfrac{79}{8}$

19. $2.75 = \dfrac{275}{100} = \dfrac{11}{4}$

20. $6.11 = \dfrac{611}{100}$

1. $\begin{array}{r} 0.22 \\ +\ 0.05 \\ \hline 0.27 \end{array}$

2. $\begin{array}{r} 0.46 \\ +\ 0.48 \\ \hline 0.94 \end{array}$

3. $\begin{array}{r} 0.53 \\ +\ 0.31 \\ \hline 0.84 \end{array}$

4. $\begin{array}{r} 0.68 \\ +\ 0.72 \\ \hline 1.4 \end{array}$

5. $\begin{array}{r} 0.42 \\ +\ 0.21 \\ \hline 0.63 \end{array}$

6. $\begin{array}{r} 0.54 \\ +\ 0.11 \\ \hline 0.65 \end{array}$

7. $\begin{array}{r} 0.48 \\ +\ 0.36 \\ \hline 0.84 \end{array}$

8. $\begin{array}{r} 0.73 \\ +\ 0.41 \\ \hline 1.14 \end{array}$

9. $\begin{array}{r} 0.79 \\ +\ 0.68 \\ \hline 1.47 \end{array}$

10. $\begin{array}{r} 0.65 \\ +\ 0.23 \\ \hline 0.88 \end{array}$

11. $\begin{array}{r} 0.66 \\ +\ 0.06 \\ \hline 0.72 \end{array}$

12. $\begin{array}{r} 0.62 \\ +\ 0.38 \\ \hline 1 \end{array}$

13. $\begin{array}{r} 0.47 \\ +\ 0.04 \\ \hline 0.51 \end{array}$

14. $\begin{array}{r} 0.39 \\ +\ 0.62 \\ \hline 1.01 \end{array}$

15. $\begin{array}{r} 0.8 \\ +\ 0.34 \\ \hline 1.14 \end{array}$

16. $\begin{array}{r} 0.07 \\ +\ 0.49 \\ \hline 0.56 \end{array}$

17. $\begin{array}{r} 0.66 \\ +\ 0.83 \\ \hline 1.49 \end{array}$

18. $\begin{array}{r} 0.59 \\ +\ 0.86 \\ \hline 1.45 \end{array}$

19. $\begin{array}{r} 0.27 \\ +\ 0.03 \\ \hline 0.3 \end{array}$

20. $\begin{array}{r} 0.76 \\ +\ 0.14 \\ \hline 0.9 \end{array}$

21. $\begin{array}{r} 0.97 \\ +\ 0.76 \\ \hline 1.73 \end{array}$

22. $\begin{array}{r} 0.49 \\ +\ 0.76 \\ \hline 1.25 \end{array}$

23. $\begin{array}{r} 0.46 \\ +\ 0.56 \\ \hline 1.02 \end{array}$

24. $\begin{array}{r} 0.91 \\ +\ 0.43 \\ \hline 1.34 \end{array}$

1. $\begin{array}{r} 0.13 \\ +\ 0.77 \\ \hline 0.9 \end{array}$

2. $\begin{array}{r} 0.58 \\ +\ 0.33 \\ \hline 0.91 \end{array}$

3. $\begin{array}{r} 0.63 \\ +\ 0.9 \\ \hline 1.53 \end{array}$

4. $\begin{array}{r} 0.7 \\ +\ 0.98 \\ \hline 1.68 \end{array}$

5. $\begin{array}{r} 0.89 \\ +\ 0.71 \\ \hline 1.6 \end{array}$

6. $\begin{array}{r} 0.14 \\ +\ 0.22 \\ \hline 0.36 \end{array}$

7. $\begin{array}{r} 0.41 \\ +\ 0.97 \\ \hline 1.38 \end{array}$

8. $\begin{array}{r} 0.83 \\ +\ 0.87 \\ \hline 1.7 \end{array}$

9. $\begin{array}{r} 0.16 \\ +\ 0.82 \\ \hline 0.98 \end{array}$

10. $\begin{array}{r} 0.57 \\ +\ 0.93 \\ \hline 1.5 \end{array}$

11. $\begin{array}{r} 0.38 \\ +\ 0.02 \\ \hline 0.4 \end{array}$

12. $\begin{array}{r} 0.33 \\ +\ 0.78 \\ \hline 1.11 \end{array}$

13. $\begin{array}{r} 0.86 \\ +\ 0.93 \\ \hline 1.79 \end{array}$

14. $\begin{array}{r} 0.41 \\ +\ 0.86 \\ \hline 1.27 \end{array}$

15. $\begin{array}{r} 0.99 \\ +\ 0.7 \\ \hline 1.69 \end{array}$

16. $\begin{array}{r} 0.27 \\ +\ 0.23 \\ \hline 0.5 \end{array}$

17. $\begin{array}{r} 0.8 \\ +\ 0.37 \\ \hline 1.17 \end{array}$

18. $\begin{array}{r} 0.3 \\ +\ 0.12 \\ \hline 0.42 \end{array}$

19. $\begin{array}{r} 0.29 \\ +\ 0.79 \\ \hline 1.08 \end{array}$

20. $\begin{array}{r} 0.11 \\ +\ 0.06 \\ \hline 0.17 \end{array}$

21. $\begin{array}{r} 0.66 \\ +\ 0.66 \\ \hline 1.32 \end{array}$

22. $\begin{array}{r} 0.88 \\ +\ 0.2 \\ \hline 1.08 \end{array}$

23. $\begin{array}{r} 0.9 \\ +\ 0.91 \\ \hline 1.81 \end{array}$

24. $\begin{array}{r} 0.74 \\ +\ 0.14 \\ \hline 0.88 \end{array}$

1. $\begin{array}{r} 2.37 \\ +\ 1.22 \\ \hline 3.59 \end{array}$

2. $\begin{array}{r} 1.45 \\ +\ 0.91 \\ \hline 2.36 \end{array}$

3. $\begin{array}{r} 2.21 \\ +\ 1.05 \\ \hline 3.26 \end{array}$

4. $\begin{array}{r} 0.5 \\ +\ 2.31 \\ \hline 2.81 \end{array}$

5. $\begin{array}{r} 1.76 \\ +\ 0.51 \\ \hline 2.27 \end{array}$

6. $\begin{array}{r} 2.49 \\ +\ 1.39 \\ \hline 3.88 \end{array}$

7. $\begin{array}{r} 0.39 \\ +\ 2.88 \\ \hline 3.27 \end{array}$

8. $\begin{array}{r} 1.37 \\ +\ 1.1 \\ \hline 2.47 \end{array}$

9. $\begin{array}{r} 2.68 \\ +\ 1.78 \\ \hline 4.46 \end{array}$

10. $\begin{array}{r} 2.89 \\ +\ 2.69 \\ \hline 5.58 \end{array}$

11. $\begin{array}{r} 1.77 \\ +\ 1.02 \\ \hline 2.79 \end{array}$

12. $\begin{array}{r} 1.05 \\ +\ 0.97 \\ \hline 2.02 \end{array}$

13. $\begin{array}{r} 0.17 \\ +\ 2.27 \\ \hline 2.44 \end{array}$

14. $\begin{array}{r} 0.23 \\ +\ 2.98 \\ \hline 3.21 \end{array}$

15. $\begin{array}{r} 2.33 \\ +\ 0.37 \\ \hline 2.7 \end{array}$

16. $\begin{array}{r} 2.08 \\ +\ 2.15 \\ \hline 4.23 \end{array}$

17. $\begin{array}{r} 2.8 \\ +\ 2.04 \\ \hline 4.84 \end{array}$

18. $\begin{array}{r} 1.71 \\ +\ 1.41 \\ \hline 3.12 \end{array}$

19. $\begin{array}{r} 2.44 \\ +\ 2.84 \\ \hline 5.28 \end{array}$

20. $\begin{array}{r} 1.57 \\ +\ 0.98 \\ \hline 2.55 \end{array}$

21. $\begin{array}{r} 2.94 \\ +\ 0.77 \\ \hline 3.71 \end{array}$

22. $\begin{array}{r} 0.51 \\ +\ 2.84 \\ \hline 3.35 \end{array}$

23. $\begin{array}{r} 0.4 \\ +\ 1.15 \\ \hline 1.55 \end{array}$

24. $\begin{array}{r} 1.23 \\ +\ 0.54 \\ \hline 1.77 \end{array}$

1. $\begin{array}{r} 2.3 \\ +\ 1.81 \\ \hline 4.11 \end{array}$

2. $\begin{array}{r} 2.65 \\ +\ 2.8 \\ \hline 5.45 \end{array}$

3. $\begin{array}{r} 1.95 \\ +\ 3.44 \\ \hline 5.39 \end{array}$

4. $\begin{array}{r} 3.29 \\ +\ 0.82 \\ \hline 4.11 \end{array}$

5. $\begin{array}{r} 2.12 \\ +\ 0.39 \\ \hline 2.51 \end{array}$

6. $\begin{array}{r} 2.25 \\ +\ 0.98 \\ \hline 3.23 \end{array}$

7. $\begin{array}{r} 2.1 \\ +\ 1.43 \\ \hline 3.53 \end{array}$

8. $\begin{array}{r} 2.11 \\ +\ 1.74 \\ \hline 3.85 \end{array}$

9. $\begin{array}{r} 0.74 \\ +\ 0.16 \\ \hline 0.9 \end{array}$

10. $\begin{array}{r} 0.53 \\ +\ 2.18 \\ \hline 2.71 \end{array}$

11. $\begin{array}{r} 0.17 \\ +\ 3.09 \\ \hline 3.26 \end{array}$

12. $\begin{array}{r} 3.71 \\ +\ 2.61 \\ \hline 6.32 \end{array}$

13. $\begin{array}{r} 2.02 \\ +\ 3.25 \\ \hline 5.27 \end{array}$

14. $\begin{array}{r} 1.41 \\ +\ 2.12 \\ \hline 3.53 \end{array}$

15. $\begin{array}{r} 3.04 \\ +\ 3.9 \\ \hline 6.94 \end{array}$

16. $\begin{array}{r} 1.14 \\ +\ 1.71 \\ \hline 2.85 \end{array}$

17. $\begin{array}{r} 1.62 \\ +\ 2.99 \\ \hline 4.61 \end{array}$

18. $\begin{array}{r} 3.87 \\ +\ 3.9 \\ \hline 7.77 \end{array}$

19. $\begin{array}{r} 2.48 \\ +\ 3.16 \\ \hline 5.64 \end{array}$

20. $\begin{array}{r} 0.87 \\ +\ 2.81 \\ \hline 3.68 \end{array}$

21. $\begin{array}{r} 3.57 \\ +\ 3.89 \\ \hline 7.46 \end{array}$

22. $\begin{array}{r} 2.71 \\ +\ 1.06 \\ \hline 3.77 \end{array}$

23. $\begin{array}{r} 3.38 \\ +\ 2.6 \\ \hline 5.98 \end{array}$

24. $\begin{array}{r} 2.97 \\ +\ 1.41 \\ \hline 4.38 \end{array}$

Lesson 9-5   Adding decimals

Name: _____

1. 1.58 + 4.71 = 6.29
2. 1.11 + 1.32 = 2.43
3. 2.56 + 1.92 = 4.48
4. 4.47 + 2.74 = 7.21
5. 0.73 + 1.19 = 1.92
6. 0.91 + 1.77 = 2.68
7. 2.59 + 1.77 = 4.36
8. 2.32 + 2.26 = 4.58
9. 4.21 + 4.61 = 8.82
10. 4.98 + 1.19 = 6.17
11. 3.26 + 4.05 = 7.31
12. 1.64 + 1.83 = 3.47
13. 3.06 + 2.69 = 5.75
14. 3.47 + 2.21 = 5.68
15. 3.07 + 2.96 = 6.03
16. 1.66 + 0.87 = 2.53
17. 2.05 + 0.75 = 2.8
18. 0.29 + 2.91 = 3.2
19. 3.54 + 3.9 = 7.44
20. 3.44 + 1.31 = 4.75
21. 2.47 + 3.89 = 6.36
22. 2.78 + 4.14 = 6.92
23. 0.2 + 3.15 = 3.35
24. 2.7 + 3.9 = 6.6

Lesson 9-6   Adding decimals

Name: _____

1. 2.23 + 2.76 = 4.99
2. 1.42 + 4.25 = 5.67
3. 4.7 + 1.47 = 6.17
4. 1.51 + 4.75 = 6.26
5. 5.65 + 4.03 = 9.68
6. 4.24 + 5.21 = 9.45
7. 3.49 + 3.88 = 7.37
8. 0.99 + 1.29 = 2.28
9. 5.76 + 0.95 = 6.71
10. 3.82 + 2.24 = 6.06
11. 1.53 + 2.82 = 4.35
12. 5.16 + 3.22 = 8.38
13. 0.5 + 2.02 = 2.52
14. 3.96 + 5.78 = 9.74
15. 1.04 + 1.69 = 2.73
16. 3.14 + 0.67 = 3.81
17. 3.64 + 5.87 = 9.51
18. 5.21 + 2.14 = 7.35
19. 1.44 + 4.75 = 6.19
20. 4.2 + 1.35 = 5.55
21. 5.98 + 2.4 = 8.38
22. 4.5 + 3.92 = 8.42
23. 4.22 + 4.19 = 8.41
24. 4.99 + 2.91 = 7.9

Lesson 9-7   Adding decimals

Name: _____

1. 2.21 + 3.9 = 6.11
2. 2.53 + 1.33 = 3.86
3. 5.22 + 3.69 = 8.91
4. 3.47 + 0.61 = 4.08
5. 5.5 + 2.51 = 8.01
6. 3.34 + 1.51 = 4.85
7. 0.81 + 1.32 = 2.13
8. 1.88 + 5.87 = 7.75
9. 2.41 + 0.8 = 3.21
10. 1.77 + 2.08 = 3.85
11. 5.78 + 1.88 = 7.66
12. 5.34 + 2.44 = 7.78
13. 1.9 + 3.94 = 5.84
14. 1.15 + 5.98 = 7.13
15. 1.28 + 4.66 = 5.94
16. 3.08 + 5.08 = 8.16
17. 0.88 + 2.57 = 3.45
18. 4.04 + 2.48 = 6.52
19. 3.04 + 4.4 = 7.44
20. 4.14 + 3.26 = 7.4
21. 1.89 + 2.39 = 4.28
22. 2.83 + 5.88 = 8.71
23. 3.97 + 1.24 = 5.21
24. 4.68 + 4.54 = 9.22

Lesson 9-8   Adding decimals

Name: _____

1. 0.91 + 1.3 = 2.21
2. 4.12 + 3.49 = 7.61
3. 3.7 + 1.86 = 5.56
4. 0.57 + 4.62 = 5.19
5. 1.94 + 4.97 = 6.91
6. 3.83 + 2.43 = 6.26
7. 4.9 + 2.78 = 7.68
8. 2.18 + 5.86 = 8.04
9. 1.87 + 2.33 = 4.2
10. 2.98 + 4.07 = 7.05
11. 3.7 + 1.51 = 5.21
12. 2.26 + 4.92 = 7.18
13. 5.17 + 3.43 = 8.6
14. 2.95 + 4.2 = 7.15
15. 1.35 + 3.07 = 4.42
16. 3.76 + 0.64 = 4.4
17. 4.44 + 1.31 = 5.75
18. 3.46 + 1.55 = 5.01
19. 3.65 + 3.37 = 7.02
20. 1.76 + 4.42 = 6.18
21. 4.7 + 5.01 = 9.71
22. 2.23 + 4.02 = 6.25
23. 4.3 + 1.42 = 5.72
24. 1.92 + 3.54 = 5.46

1. 6.24 + 2.56 = 8.8
2. 5.14 + 2.72 = 7.86
3. 3.55 + 5.34 = 8.89
4. 5.75 + 5.76 = 11.51
5. 2.13 + 6.79 = 8.92
6. 6.85 + 1.86 = 8.71
7. 6.25 + 2.5 = 8.75
8. 5.83 + 6.18 = 12.01
9. 5.41 + 2.4 = 7.81
10. 1.19 + 4.59 = 5.78
11. 3.87 + 3.84 = 7.71
12. 1.67 + 5.1 = 6.77
13. 6.08 + 1.26 = 7.34
14. 5.56 + 5.25 = 10.81
15. 0.93 + 3.6 = 4.53
16. 1.4 + 2.47 = 3.87
17. 1.54 + 3.2 = 4.74
18. 1.45 + 3.38 = 4.83
19. 2.46 + 1.35 = 3.81
20. 1.44 + 5.04 = 6.48
21. 1.67 + 5.94 = 7.61
22. 3.55 + 6.52 = 10.07
23. 6.53 + 1.36 = 7.89
24. 2.35 + 1.21 = 3.56

1. 4.64 + 5.1 = 9.74
2. 3.85 + 5.26 = 9.11
3. 4.88 + 7.51 = 12.39
4. 2.68 + 4.09 = 6.77
5. 1.01 + 2.64 = 3.65
6. 1.5 + 2.28 = 3.78
7. 2.8 + 1.53 = 4.33
8. 3.2 + 1.19 = 4.39
9. 6.56 + 7.26 = 13.82
10. 7.46 + 4.75 = 12.21
11. 4.2 + 2.68 = 6.88
12. 7.12 + 1.36 = 8.48
13. 5.28 + 5.15 = 10.43
14. 7.12 + 3.6 = 10.72
15. 6.16 + 2.92 = 9.08
16. 1.48 + 4.43 = 5.91
17. 4.65 + 2.37 = 7.02
18. 3.95 + 4.61 = 8.56
19. 5.46 + 6.3 = 11.76
20. 2.97 + 7.41 = 10.38
21. 4.24 + 4.37 = 8.61
22. 3.98 + 2.84 = 6.82
23. 7.43 + 6.72 = 14.15
24. 7.1 + 5.87 = 12.97

1. 2.70 − 2.09 = 0.61
2. 1.86 − 0.46 = 1.4
3. 2.97 − 0.44 = 2.53
4. 2.91 − 2.31 = 0.6
5. 2.65 − 0.2 = 2.45
6. 1.24 − 0.47 = 0.77
7. 0.49 − 0.33 = 0.16
8. 1.63 − 0.9 = 0.73
9. 2.05 − 1.92 = 0.13
10. 2.97 − 0.33 = 2.64
11. 1.06 − 0.82 = 0.24
12. 2.21 − 1.03 = 1.18
13. 2.06 − 0.86 = 1.2
14. 0.32 − 0.19 = 0.13
15. 1.26 − 0.73 = 0.53
16. 1.08 − 0.37 = 0.71
17. 2.51 − 0.98 = 1.53
18. 1.87 − 0.42 = 1.45
19. 2.59 − 1.2 = 1.39
20. 1.47 − 0.51 = 0.96
21. 1.85 − 1.1 = 0.75
22. 2.80 − 1.16 = 1.64
23. 2.73 − 0.55 = 2.18
24. 1.99 − 0.92 = 1.07

1. 2.85 − 0.45 = 2.4
2. 2.21 − 1.8 = 0.41
3. 1.55 − 0.39 = 1.16
4. 2.21 − 2.05 = 0.16
5. 2.51 − 0.84 = 1.67
6. 1.64 − 0.09 = 1.55
7. 1.32 − 0.65 = 0.67
8. 2.56 − 0.7 = 1.86
9. 1.65 − 0.44 = 1.21
10. 2.96 − 2.08 = 0.88
11. 2.47 − 1.25 = 1.22
12. 2.36 − 1.99 = 0.37
13. 2.92 − 0.13 = 2.79
14. 2.16 − 1.02 = 1.14
15. 2.00 − 1.41 = 0.59
16. 2.01 − 0.26 = 1.75
17. 2.23 − 0.96 = 1.27
18. 1.40 − 1.17 = 0.23
19. 1.28 − 0.29 = 0.99
20. 1.94 − 0.71 = 1.23
21. 2.49 − 0.8 = 1.69
22. 2.03 − 0.39 = 1.64
23. 1.72 − 0.25 = 1.47
24. 3.00 − 1.37 = 1.63

1.
```
  2.3 5
- 1.6 2
-------
  0.7 3
```
2.
```
  1.2 3
- 0.0 4
-------
  1.1 9
```
3.
```
  2.5 6
- 2.0 7
-------
  0.4 9
```
4.
```
  1.5 1
- 0.0 8
-------
  1.4 3
```

5.
```
  2.3 7
- 0.2 8
-------
  2.0 9
```
6.
```
  1.9 6
- 1.4 6
-------
  0.5
```
7.
```
  2.6 5
- 1.6 1
-------
  1.0 4
```
8.
```
  2.4 1
- 1.7 4
-------
  0.6 7
```

9.
```
  1.7 9
- 0.4 7
-------
  1.3 2
```
10.
```
  2.0 8
- 1.2 7
-------
  0.8 1
```
11.
```
  1.3 4
- 1.0 2
-------
  0.3 2
```
12.
```
  2.9 4
- 0.1 5
-------
  2.7 9
```

13.
```
  1.1 5
- 0.5 2
-------
  0.6 3
```
14.
```
  1.7 3
- 1.3 3
-------
  0.4
```
15.
```
  2.7 6
- 1.9 8
-------
  0.7 8
```
16.
```
  2.0 0
- 0.4 2
-------
  1.5 8
```

17.
```
  2.4 1
- 1.2 6
-------
  1.1 5
```
18.
```
  1.7 6
- 0.8 4
-------
  0.9 2
```
19.
```
  2.8 6
- 2.6 5
-------
  0.2 1
```
20.
```
  2.6 0
- 1.6 1
-------
  0.9 9
```

21.
```
  2.6 4
- 1
-------
  1.6 4
```
22.
```
  2.5 8
- 1.3 2
-------
  1.2 6
```
23.
```
  1.7 6
- 0.3 9
-------
  1.3 7
```
24.
```
  2.1 7
- 0.4 8
-------
  1.6 9
```

1.
```
  2.9 1
- 2.2 1
-------
  0.7
```
2.
```
  2.0 4
- 1.6 4
-------
  0.4
```
3.
```
  1.2 3
- 0.8 5
-------
  0.3 8
```
4.
```
  2.6 5
- 0.4 3
-------
  2.2 2
```

5.
```
  3.8 3
- 1.8 9
-------
  1.9 4
```
6.
```
  2.3 3
- 1.5
-------
  0.8 3
```
7.
```
  1.8 9
- 1.5 2
-------
  0.3 7
```
8.
```
  3.7 9
- 0.8 7
-------
  2.9 2
```

9.
```
  3.0 5
- 2.4 6
-------
  0.5 9
```
10.
```
  2.1 6
- 1.6 5
-------
  0.5 1
```
11.
```
  2.0 3
- 2.0 2
-------
  0.0 1
```
12.
```
  3.8 4
- 1.5 3
-------
  2.3 1
```

13.
```
  2.7 0
- 1.4 1
-------
  1.2 9
```
14.
```
  3.6 8
- 3.2 8
-------
  0.4
```
15.
```
  1.2 0
- 0.2 7
-------
  0.9 3
```
16.
```
  3.1 2
- 0.4 8
-------
  2.6 4
```

17.
```
  3.2 5
- 3
-------
  0.2 5
```
18.
```
  3.1 9
- 2.0 6
-------
  1.1 3
```
19.
```
  3.0 7
- 1.8 2
-------
  1.2 5
```
20.
```
  2.4 1
- 1.1 5
-------
  1.2 6
```

21.
```
  1.9 4
- 0.5 5
-------
  1.3 9
```
22.
```
  2.5 9
- 0.8 1
-------
  1.7 8
```
23.
```
  3.6 2
- 2.4 6
-------
  1.1 6
```
24.
```
  2.0 7
- 0.7 9
-------
  1.2 8
```

1.
```
  3.4 1
- 2.8 4
-------
  0.5 7
```
2.
```
  3.1 7
- 2.3 7
-------
  0.8
```
3.
```
  3.5 0
- 0.5 8
-------
  2.9 2
```
4.
```
  2.4 2
- 0.4 8
-------
  1.9 4
```

5.
```
  2.5 1
- 1.5 7
-------
  0.9 4
```
6.
```
  2.5 8
- 1.7
-------
  0.8 8
```
7.
```
  2.8 1
- 1.8 6
-------
  0.9 5
```
8.
```
  2.8 6
- 1.7 8
-------
  1.0 8
```

9.
```
  3.0 4
- 1.0 8
-------
  1.9 6
```
10.
```
  3.3 4
- 0.6 1
-------
  2.7 3
```
11.
```
  1.4 7
- 0.5 7
-------
  0.9
```
12.
```
  2.1 5
- 0.4 2
-------
  1.7 3
```

13.
```
  2.0 6
- 1.9
-------
  0.1 6
```
14.
```
  1.8 7
- 0.1 7
-------
  1.7
```
15.
```
  1.4 0
- 0.7 9
-------
  0.6 1
```
16.
```
  3.4 0
- 2.4 6
-------
  0.9 4
```

17.
```
  2.0 4
- 0.5 8
-------
  1.4 6
```
18.
```
  2.7 0
- 2.0 5
-------
  0.6 5
```
19.
```
  3.0 0 0
- 0.1 6 8
---------
  2.8 3 2
```
20.
```
  3.7 8
- 2.5 4
-------
  1.2 4
```

21.
```
  3.3 2
- 1.6 4
-------
  1.6 8
```
22.
```
  3.1 1
- 1.9 5
-------
  1.1 6
```
23.
```
  2.6 8
- 1.7 5
-------
  0.9 3
```
24.
```
  1.8 0
- 0.5 5
-------
  1.2 5
```

1.
```
  2.9 0
- 1.1 4
-------
  1.7 6
```
2.
```
  3.3 3
- 2.0 9
-------
  1.2 4
```
3.
```
  3.7 0
- 1.9 6
-------
  1.7 4
```
4.
```
  3.5 1
- 1.2 2
-------
  2.2 9
```

5.
```
  2.5 7
- 1.7 8
-------
  0.7 9
```
6.
```
  2.9 3
- 1.1 6
-------
  1.7 7
```
7.
```
  2.0 6
- 0.3 9
-------
  1.6 7
```
8.
```
  3.2 1
- 0.3 4
-------
  2.8 7
```

9.
```
  3.8 5
- 1.5 2
-------
  2.3 3
```
10.
```
  2.0 7
- 1.9 3
-------
  0.1 4
```
11.
```
  4.0 0
- 3.1 3
-------
  0.8 7
```
12.
```
  3.6 2
- 3.1 6
-------
  0.4 6
```

13.
```
  3.6 0
- 1.9 2
-------
  1.6 8
```
14.
```
  1.1 5
- 0.1 6
-------
  0.9 9
```
15.
```
  2.9 5
- 1
-------
  1.9 5
```
16.
```
  3.7 0
- 0.3 8
-------
  3.3 2
```

17.
```
  2.8 5
- 0.9
-------
  1.9 5
```
18.
```
  1.8 6
- 1.5 9
-------
  0.2 7
```
19.
```
  1.3 4
- 0.6 6
-------
  0.6 8
```
20.
```
  1.7 8
- 1.1 1
-------
  0.6 7
```

21.
```
  3.0 7
- 1.2 6
-------
  1.8 1
```
22.
```
  2.6 1
- 0.7 5
-------
  1.8 6
```
23.
```
  3.5 3
- 2.9 4
-------
  0.5 9
```
24.
```
  3.3 4
- 1.6 9
-------
  1.6 5
```

1. 3.12 − 0.27 = 2.85
2. 4.83 − 0.89 = 3.94
3. 3.18 − 0.88 = 2.3
4. 2.61 − 2.28 = 0.33

5. 4.88 − 2.55 = 2.33
6. 3.78 − 2.96 = 0.82
7. 4.46 − 4.07 = 0.39
8. 3.82 − 2.64 = 1.18

9. 1.54 − 0.46 = 1.08
10. 4.00 − 2.98 = 1.02
11. 1.71 − 0.98 = 0.73
12. 4.64 − 0.36 = 4.28

13. 3.66 − 2.1 = 1.56
14. 4.96 − 1.76 = 3.2
15. 2.26 − 2 = 0.26
16. 3.91 − 1.12 = 2.79

17. 2.53 − 1.13 = 1.4
18. 2.96 − 2.79 = 0.17
19. 2.75 − 1.61 = 1.14
20. 4.98 − 4.11 = 0.87

21. 4.80 − 2.81 = 1.99
22. 3.94 − 2.76 = 1.18
23. 4.29 − 1.6 = 2.69
24. 4.22 − 0.25 = 3.97

1. 4.72 − 3.57 = 1.15
2. 2.18 − 1.15 = 1.03
3. 4.72 − 0.14 = 4.58
4. 3.60 − 1.84 = 1.76

5. 4.83 − 1.1 = 3.73
6. 2.24 − 1.9 = 0.34
7. 4.06 − 1.36 = 2.7
8. 4.77 − 0.15 = 4.62

9. 3.08 − 0.63 = 2.45
10. 3.79 − 2 = 1.79
11. 3.41 − 0.74 = 2.67
12. 4.80 − 3.37 = 1.43

13. 4.61 − 3.17 = 1.44
14. 4.55 − 1.27 = 3.28
15. 4.21 − 3.3 = 0.91
16. 4.12 − 3.43 = 0.69

17. 3.29 − 2.11 = 1.18
18. 3.93 − 1.6 = 2.33
19. 4.99 − 4.01 = 0.98
20. 2.93 − 2.15 = 0.78

21. 3.28 − 0.62 = 2.66
22. 4.87 − 4.26 = 0.61
23. 4.42 − 3.98 = 0.44
24. 3.76 − 1.87 = 1.89

1. 3.96 − 1.91 = 2.05
2. 1.80 − 0.77 = 1.03
3. 5.27 − 1.84 = 3.43
4. 2.48 − 2.11 = 0.37

5. 2.57 − 2.14 = 0.43
6. 1.88 − 1.51 = 0.37
7. 2.11 − 0.4 = 1.71
8. 4.23 − 0.49 = 3.74

9. 5.52 − 4.32 = 1.2
10. 2.19 − 1.23 = 0.96
11. 5.90 − 3.17 = 2.73
12. 2.50 − 1.53 = 0.97

13. 3.17 − 1.09 = 2.08
14. 3.27 − 1.51 = 1.76
15. 5.03 − 4.62 = 0.41
16. 5.26 − 3.31 = 1.95

17. 2.28 − 1.92 = 0.36
18. 4.15 − 2.99 = 1.16
19. 2.73 − 1.03 = 1.7
20. 3.84 − 2.03 = 1.81

21. 2.36 − 1.07 = 1.29
22. 5.47 − 3.62 = 1.85
23. 4.89 − 3.94 = 0.95
24. 5.72 − 2.07 = 3.65

1. 3.38 − 2.09 = 1.29
2. 5.64 − 2.74 = 2.9
3. 2.91 − 1 = 1.91
4. 5.35 − 1.77 = 3.58

5. 6.27 − 4.44 = 1.83
6. 3.16 − 2.85 = 0.31
7. 3.66 − 0.89 = 2.77
8. 5.06 − 2.22 = 2.84

9. 4.16 − 0.45 = 3.71
10. 4.38 − 3.97 = 0.41
11. 6.58 − 5.97 = 0.61
12. 6.97 − 4.83 = 2.14

13. 6.51 − 1.04 = 5.47
14. 5.16 − 3.89 = 1.27
15. 4.24 − 3.2 = 1.04
16. 1.52 − 1.08 = 0.44

17. 4.88 − 3.32 = 1.56
18. 6.24 − 5.83 = 0.41
19. 5.33 − 4.64 = 0.69
20. 6.07 − 3.92 = 2.15

21. 6.95 − 2.73 = 4.22
22. 3.32 − 1.76 = 1.56
23. 2.71 − 1.27 = 1.44
24. 4.99 − 0.53 = 4.46

1.
```
×  0.4 2
      0.7
 ───────
    2 9 4
 ───────
  0.2 9 4
```

2.
```
×  0.9 5
      0.8
 ───────
    7 6 0
 ───────
  0.7 6 0
```

3.
```
×  0.5 4
      0.5
 ───────
    2 7 0
 ───────
  0.2 7 0
```

4.
```
×  0.4 9
      0.3
 ───────
    1 4 7
 ───────
  0.1 4 7
```

5.
```
×  0.6 1
      0.9
 ───────
    5 4 9
 ───────
  0.5 4 9
```

6.
```
×  0.8 2
      0.4
 ───────
    3 2 8
 ───────
  0.3 2 8
```

7.
```
×  0.1 8
      0.7
 ───────
    1 2 6
 ───────
  0.1 2 6
```

8.
```
×  0.9 3
      0.5
 ───────
    4 6 5
 ───────
  0.4 6 5
```

9.
```
×  0.7 4
      0.9
 ───────
    6 6 6
 ───────
  0.6 6 6
```

10.
```
×  0.4 3
      0.5
 ───────
    2 1 5
 ───────
  0.2 1 5
```

11.
```
×  0.1 9
      0.9
 ───────
    1 7 1
 ───────
  0.1 7 1
```

12.
```
×  0.4 4
      0.6
 ───────
    2 6 4
 ───────
  0.2 6 4
```

13.
```
×  0.5 3
      0.4
 ───────
    2 1 2
 ───────
  0.2 1 2
```

14.
```
×  0.2 9
      0.3
 ───────
    0 8 7
 ───────
  0.0 8 7
```

15.
```
×  0.4 9
      0.3
 ───────
    1 4 7
 ───────
  0.1 4 7
```

16.
```
×  0.7 2
      0.2
 ───────
    1 4 4
 ───────
  0.1 4 4
```

17.
```
×  0.9 4
      0.2
 ───────
    1 8 8
 ───────
  0.1 8 8
```

18.
```
×  0.8 4
      0.7
 ───────
    5 8 8
 ───────
  0.5 8 8
```

19.
```
×  0.7 3
      0.8
 ───────
    5 8 4
 ───────
  0.5 8 4
```

20.
```
×  0.9 2
      0.6
 ───────
    5 5 2
 ───────
  0.5 5 2
```

1.
```
×   2.5
    6.3
 ───────
    7 5
  1 5 0
 ───────
 1 5.7 5
```

2.
```
×   5.2
    1.2
 ───────
  1 0 4
    5 2
 ───────
  6.2 4
```

3.
```
×   5.2
    2.7
 ───────
  3 6 4
  1 0 4
 ───────
 1 4.0 4
```

4.
```
×   7.6
    4.1
 ───────
    7 6
  3 0 4
 ───────
 3 1.1 6
```

5.
```
×   8.5
    3.3
 ───────
  2 5 5
  2 5 5
 ───────
 2 8.0 5
```

6.
```
×   6.5
    2.6
 ───────
  3 9 0
  1 3 0
 ───────
 1 6.9 0
```

7.
```
×   5.6
    3.6
 ───────
  3 3 6
  1 6 8
 ───────
 2 0.1 6
```

8.
```
×   4.3
    9.4
 ───────
  1 7 2
  3 8 7
 ───────
 4 0.4 2
```

9.
```
×   7.4
    1.4
 ───────
  2 9 6
    7 4
 ───────
 1 0.3 6
```

10.
```
×   3.4
    9.9
 ───────
  3 0 6
  3 0 6
 ───────
 3 3.6 6
```

11.
```
×   9.6
    4.9
 ───────
  8 6 4
  3 8 4
 ───────
 4 7.0 4
```

12.
```
×   2.1
    2.1
 ───────
    2 1
    4 2
 ───────
  4.4 1
```

1.
```
×   1.4
    1.7
 ───────
    9 8
    1 4
 ───────
  2.3 8
```

2.
```
×   2.4
    7.2
 ───────
    4 8
  1 6 8
 ───────
 1 7.2 8
```

3.
```
×   9.5
    9.3
 ───────
  2 8 5
  8 5 5
 ───────
 8 8.3 5
```

4.
```
×   6.2
    1.8
 ───────
  4 9 6
    6 2
 ───────
 1 1.1 6
```

5.
```
×   4.4
    2.9
 ───────
  3 9 6
    8 8
 ───────
 1 2.7 6
```

6.
```
×   9.2
    2.5
 ───────
  4 6 0
  1 8 4
 ───────
 2 3.0 0
```

7.
```
×   1.3
    9.8
 ───────
  1 0 4
  1 1 7
 ───────
 1 2.7 4
```

8.
```
×   5.3
    2.5
 ───────
  2 6 5
  1 0 6
 ───────
 1 3.2 5
```

9.
```
×   7.1
    9.4
 ───────
  2 8 4
  6 3 9
 ───────
 6 6.7 4
```

10.
```
×   3.9
    3.5
 ───────
  1 9 5
  1 1 7
 ───────
 1 3.6 5
```

11.
```
×   4.1
    5.2
 ───────
    8 2
  2 0 5
 ───────
 2 1.3 2
```

12.
```
×   3.6
    8.8
 ───────
  2 8 8
  2 8 8
 ───────
 3 1.6 8
```

1.
```
×   1.4 1
      2.8
 ─────────
  1 1 2 8
  2 8 2
 ─────────
  3.9 4 8
```

2.
```
×   3.1 4
      3.7
 ─────────
  2 1 9 8
  9 4 2
 ─────────
 1 1.6 1 8
```

3.
```
×   4.2 5
      1.4
 ─────────
  1 7 0 0
  4 2 5
 ─────────
  5.9 5 0
```

4.
```
×   2.8 2
      3.2
 ─────────
  5 6 4
  8 4 6
 ─────────
  9.0 2 4
```

5.
```
×   4.2 7
      4.6
 ─────────
  2 5 6 2
  1 7 0 8
 ─────────
 1 9.6 4 2
```

6.
```
×   3.5 5
      3.5
 ─────────
  1 7 7 5
  1 0 6 5
 ─────────
 1 2.4 2 5
```

7.
```
×   3.3 3
      4.5
 ─────────
  1 6 6 5
  1 3 3 2
 ─────────
 1 4.9 8 5
```

8.
```
×   3.2 3
      2.1
 ─────────
    3 2 3
    6 4 6
 ─────────
  6.7 8 3
```

9.
```
×   2.2 3
      2.6
 ─────────
  1 3 3 8
    4 4 6
 ─────────
  5.7 9 8
```

10.
```
×   0.8 6
      1.3
 ─────────
  2 5 8
    8 6
 ─────────
  1.1 1 8
```

11.
```
×   4.0 1
      1.3
 ─────────
  1 2 0 3
  4 0 1
 ─────────
  5.2 1 3
```

12.
```
×   4.3 9
      1.5
 ─────────
  2 1 9 5
  4 3 9
 ─────────
  6.5 8 5
```

1.
```
×  4.1 2
      3.6
   2 4 7 2
 1 2 3 6
 1 4.8 3 2
```
2.
```
×  2.0 8
      8.2
     4 1 6
 1 6 6 4
 1 7.0 5 6
```
3.
```
×  3.5 7
      3.5
   1 7 8 5
 1 0 7 1
 1 2.4 9 5
```

4.
```
×  1.3 9
      4.8
   1 1 1 2
 5 5 6
 6.6 7 2
```
5.
```
×  2.9 2
      4.3
     8 7 6
 1 1 6 8
 1 2.5 5 6
```
6.
```
×  4.5 1
      2.7
   3 1 5 7
 9 0 2
 1 2.1 7 7
```

7.
```
×  3.1 6
      3.2
     6 3 2
 9 4 8
 1 0.1 1 2
```
8.
```
×  4.7 8
      2.6
   2 8 6 8
 9 5 6
 1 2.4 2 8
```
9.
```
×  4.7 5
      1.8
   3 8 0 0
 4 7 5
 8.5 5 0
```

10.
```
×  4.6 5
      4.8
   3 7 2 0
 1 8 6 0
 2 2.3 2 0
```
11.
```
×  3.2 4
      1.6
   1 9 4 4
 3 2 4
 5.1 8 4
```
12.
```
×  2.4 1
      1.6
   1 4 4 6
 2 4 1
 3.8 5 6
```

1.
```
×  1.9 6
      1.4
     7 8 4
 1 9 6
 2.7 4 4
```
2.
```
×  5.2 1
      2.2
   1 0 4 2
 1 0 4 2
 1 1.4 6 2
```
3.
```
×  5.8 7
      1.8
   4 6 9 6
 5 8 7
 1 0.5 6 6
```

4.
```
×  4.7 6
      1.9
   4 2 8 4
 4 7 6
 9.0 4 4
```
5.
```
×  3.4 5
      8.4
   1 3 8 0
 2 7 6 0
 2 8.9 8 0
```
6.
```
×  2.7 8
      3.1
     2 7 8
 8 3 4
 8.6 1 8
```

7.
```
×  1.9 3
      5.5
     9 6 5
 9 6 5
 1 0.6 1 5
```
8.
```
×  2.8 2
      5.4
   1 1 2 8
 1 4 1 0
 1 5.2 2 8
```
9.
```
×  3.8 7
      3.6
   2 3 2 2
 1 1 6 1
 1 3.9 3 2
```

10.
```
×  7.6 7
      6.1
     7 6 7
 4 6 0 2
 4 6.7 8 7
```
11.
```
×  6.4 3
      2.6
   3 8 5 8
 1 2 8 6
 1 6.7 1 8
```
12.
```
×  4.1 5
      5.9
   3 7 3 5
 2 0 7 5
 2 4.4 8 5
```

1.
```
×  1.7 2
      2.9 3
     5 1 6
 1 5 4 8
 3 4 4
 5.0 3 9 6
```
2.
```
×  5.7 5
      1.4 2
   1 1 5 0
 2 3 0 0
 5 7 5
 8.1 6 5 0
```
3.
```
×  2.7 4
      3.0 9
   2 4 6 6
 8 2 2 ·
 8.4 6 6 6
```

4.
```
×  1.2 8
      3.5 6
     7 6 8
 6 4 0
 3 8 4
 4.5 5 6 8
```
5.
```
×  2.2 3
      7.6 3
     6 6 9
 1 3 3 8
 1 5 6 1
 1 7.0 1 4 9
```
6.
```
×  4.4 6
      7.7 1
     4 4 6
 3 1 2 2
 3 1 2 2
 3 4.3 8 6 6
```

7.
```
×  4.8 7
      4.3 2
     9 7 4
 1 4 6 1
 1 9 4 8
 2 1.0 3 8 4
```
8.
```
×  2.6 8
      1.0 8
   2 1 4 4
 2 6 8 ·
 2.8 9 4 4
```
9.
```
×  1.8 5
      6.9 2
     3 7 0
 1 6 6 5
 1 1 1 0
 1 2.8 0 2 0
```

10.
```
×  2.2 8
      5.5 9
   2 0 5 2
 1 1 4 0
 1 1 4 0
 1 2.7 4 5 2
```
11.
```
×  7.6 2
      5.7 2
   1 5 2 4
 5 3 3 4
 3 8 1 0
 4 3.5 8 6 4
```
12.
```
×  5.5 7
      1.1 1
     5 5 7
 5 5 7
 5 5 7
 6.1 8 2 7
```

1.
```
×  2.1 1
      1.8 5
   1 0 5 5
 1 6 8 8
 2 1 1
 3.9 0 3 5
```
2.
```
×  1.1 6
      5.2 4
     4 6 4
 2 3 2
 5 8 0
 6.0 7 8 4
```
3.
```
×  4.7 9
      0.3 6
   2 8 7 4
 1 4 3 7
 1.7 2 4 4
```

4.
```
×  4.4 9
      7.6 2
     8 9 8
 2 6 9 4
 3 1 4 3
 3 4.2 1 3 8
```
5.
```
×  5.3 5
      7.4 2
   1 0 7 0
 2 1 4 0
 3 7 4 5
 3 9.6 9 7 0
```
6.
```
×  5.9 6
      2.2 5
   2 9 8 0
 1 1 9 2
 1 1 9 2
 1 3.4 1 0 0
```

7.
```
×  1.3 8
      0.7 4
     5 5 2
 9 6 6
 1.0 2 1 2
```
8.
```
×  0.8 6
      6.6 5
     4 3 0
 5 1 6
 5 1 6
 5.7 1 9 0
```
9.
```
×  2.4 2
      2.6 6
   1 4 5 2
 1 4 5 2
 4 8 4
 6.4 3 7 2
```

10.
```
×  7.0 6
      7.8 5
   3 5 3 0
 5 6 4 8
 4 9 4 2
 5 5.4 2 1 0
```
11.
```
×  3.7 6
      5.9 2
     7 5 2
 3 3 8 4
 1 8 8 0
 2 2.2 5 9 2
```
12.
```
×  0.8 1
      5.4 2
     1 6 2
 3 2 4
 4 0 5
 4.3 9 0 2
```

1.
```
×  6.0 5
   3.2 3
   1 8 1 5
 1 2 1 0
 1 8 1 5
1 9.5 4 1 5
```

2.
```
×  6.4 5
   5.8 2
   1 2 9 0
 5 1 6 0
 3 2 2 5
3 7.5 3 9 0
```

3.
```
×  2.1 2
   1.8 8
   1 6 9 6
 1 6 9 6
 2 1 2
 3.9 8 5 6
```

4.
```
×  5.3 6
   5.5 1
     5 3 6
 2 6 8 0
 2 6 8 0
2 9.5 3 3 6
```

5.
```
×  3.5 9
   1.3 5
   1 7 9 5
 1 0 7 7
 3 5 9
 4.8 4 6 5
```

6.
```
×  1.4 5
   2.0 3
     4 3 5
 2 9 0 ·
 2.9 4 3 5
```

7.
```
×  6.5 9
   7.3 1
     6 5 9
 1 9 7 7
 4 6 1 3
4 8.1 7 2 9
```

8.
```
×  3.1 8
   2.3 7
   2 2 2 6
 9 5 4
 6 3 6
 7.5 3 6 6
```

9.
```
×  3.9 7
   0.6 4
   1 5 8 8
 2 3 8 2
 2.5 4 0 8
```

10.
```
×  6.9 5
   5.1 6
   4 1 7 0
 6 9 5
 3 4 7 5
3 5.8 6 2 0
```

11.
```
×  2.7 6
   0.9 3
     8 2 8
 2 4 8 4
 2.5 6 6 8
```

12.
```
×  5.8 4
   2.2 4
   2 3 3 6
 1 1 6 8
 1 1 6 8
1 3.0 8 1 6
```

244

1.
```
×  2.5 3
   4.9 4
   1 0 1 2
 2 2 7 7
 1 0 1 2
1 2.4 9 8 2
```

2.
```
×  6.5 3
   2.5 8
   5 2 2 4
 3 2 6 5
 1 3 0 6
1 6.8 4 7 4
```

3.
```
×  1.4 6
   3.8 5
     7 3 0
 1 1 6 8
 4 3 8
 5.6 2 1 0
```

4.
```
×  2.3 4
   6.7 8
   1 8 7 2
 1 6 3 8
 1 4 0 4
1 5.8 6 5 2
```

5.
```
×  5.1 1
   4.5 8
   4 0 8 8
 2 5 5 5
 2 0 4 4
2 3.4 0 3 8
```

6.
```
×  3.5 5
   3.8 8
   2 8 4 0
 2 8 4 0
 1 0 6 5
1 3.7 7 4 0
```

7.
```
×  5.5 1
   6.0 2
   1 1 0 2
 3 3 0 6 ·
3 3.1 7 0 2
```

8.
```
×  0.9 1
   2.9 5
     4 5 5
   8 1 9
 1 8 2
 2.6 8 4 5
```

9.
```
×  1.4 4
   5.5 3
     4 3 2
   7 2 0
 7 2 0
 7.9 6 3 2
```

10.
```
×  4.1 6
   3.6 9
   3 7 4 4
 2 4 9 6
 1 2 4 8
1 5.3 5 0 4
```

11.
```
×  1.1 7
   3.0 4
     4 6 8
 3 5 1 ·
 3.5 5 6 8
```

12.
```
×  3.0 7
   2.9 9
   2 7 6 3
 2 7 6 3
 6 1 4
 9.1 7 9 3
```

245

1. $7.86 \div 2 = 3.93$ 2. $9.32 \div 2 = 4.66$ 3. $6.32 \div 2 = 3.16$ 4. $9.12 \div 2 = 4.56$

5. $4.35 \div 3 = 1.45$ 6. $9.54 \div 3 = 3.18$ 7. $5.55 \div 3 = 1.85$ 8. $10.44 \div 3 = 3.48$

9. $11.12 \div 4 = 2.78$ 10. $11.76 \div 4 = 2.94$ 11. $9.36 \div 4 = 2.34$ 12. $10.84 \div 4 = 2.71$

13. $17.2 \div 5 = 3.44$ 14. $18.65 \div 5 = 3.73$ 15. $23.1 \div 5 = 4.62$ 16. $7.7 \div 5 = 1.54$

246

1. $9.28 \div 2 = 4.64$ 2. $11.26 \div 2 = 5.63$ 3. $8.58 \div 2 = 4.29$ 4. $7.14 \div 2 = 3.57$

5. $4.08 \div 3 = 1.36$ 6. $11.04 \div 3 = 3.68$ 7. $6.81 \div 3 = 2.27$ 8. $8.01 \div 3 = 2.68$

9. $9.04 \div 4 = 2.26$ 10. $10.64 \div 4 = 2.66$ 11. $6.24 \div 4 = 1.56$ 12. $15.32 \div 4 = 3.83$

13. $14.05 \div 5 = 2.81$ 14. $8.95 \div 5 = 1.79$ 15. $12.35 \div 5 = 2.47$ 16. $23.15 \div 5 = 4.63$

247

1. $7.68 \div 6 = 1.28$ 2. $16.38 \div 6 = 2.73$ 3. $21.18 \div 6 = 3.53$ 4. $29.1 \div 6 = 4.85$

5. $13.44 \div 7 = 1.92$ 6. $15.96 \div 7 = 2.28$ 7. $13.93 \div 7 = 1.99$ 8. $25.69 \div 7 = 3.67$

9. $20.61 \div 8 = 2.58$ 10. $19.92 \div 8 = 2.49$ 11. $27.76 \div 8 = 3.47$ 12. $22.88 \div 8 = 2.86$

13. $31.14 \div 9 = 3.46$ 14. $28.62 \div 9 = 3.18$ 15. $40.77 \div 9 = 4.53$ 16. $14.31 \div 9 = 1.59$

1. $21.42 \div 6 = 3.57$ 2. $29.34 \div 6 = 4.89$ 3. $28.08 \div 6 = 4.68$ 4. $10.74 \div 6 = 1.79$

5. $12.39 \div 7 = 1.77$ 6. $26.11 \div 7 = 3.73$ 7. $10.01 \div 7 = 1.43$ 8. $31.99 \div 7 = 4.57$

9. $22.64 \div 8 = 2.83$ 10. $18.88 \div 8 = 2.36$ 11. $26.32 \div 8 = 3.29$ 12. $20.08 \div 8 = 2.51$

13. $35.28 \div 9 = 3.92$ 14. $31.23 \div 9 = 3.47$ 15. $42.21 \div 9 = 4.69$ 16. $17.91 \div 9 = 1.99$

1. $7.56 \div 2.1 = 3.6$ 2. $2.38 \div 1.7 = 1.4$ 3. $8.99 \div 2.9 = 3.1$ 4. $6.48 \div 3.6 = 1.8$

5. $6.65 \div 1.9 = 3.5$ 6. $9.52 \div 3.4 = 2.8$ 7. $5.46 \div 2.1 = 2.6$ 8. $7.74 \div 1.8 = 4.3$

9. $2.88 \div 1.8 = 1.6$ 10. $5.32 \div 1.4 = 3.8$ 11. $12.21 \div 3.3 = 3.7$ 12. $10.14 \div 2.6 = 3.9$

13. $9.18 \div 3.4 = 2.7$ 14. $4.93 \div 2.9 = 1.7$ 15. $10.56 \div 2.4 = 4.4$ 16. $6.44 \div 4.6 = 1.4$

1. $2.25 \div 1.5 = 1.5$ 2. $8.82 \div 1.8 = 4.9$ 3. $8.16 \div 2.4 = 3.4$ 4. $7.02 \div 2.7 = 2.6$

5. $3.04 \div 1.9 = 1.6$ 6. $12.87 \div 3.3 = 3.9$ 7. $12.58 \div 3.7 = 3.4$ 8. $13.95 \div 3.1 = 4.5$

9. $15.17 \div 3.7 = 4.1$ 10. $14.08 \div 4.4 = 3.2$ 11. $9.31 \div 1.9 = 4.9$ 12. $18.49 \div 4.3 = 4.3$

13. $6.96 \div 2.4 = 2.9$ 14. $3.42 \div 1.9 = 1.8$ 15. $16.65 \div 4.5 = 3.7$ 16. $12.69 \div 4.7 = 2.7$

1.
$16.32 \div 3.4 = 4.8$

2.
$12.48 \div 3.2 = 3.9$

3.
$15.17 \div 4.1 = 3.7$

4.
$22.56 \div 4.7 = 4.8$

5.
$10.64 \div 1.9 = 5.6$

6.
$15.75 \div 4.5 = 3.5$

7.
$11.31 \div 2.9 = 3.9$

8.
$14.62 \div 3.4 = 4.3$

9.
$12.69 \div 4.7 = 2.7$

10.
$8.16 \div 4.8 = 1.7$

11.
$19.68 \div 4.8 = 4.1$

12.
$15.12 \div 3.6 = 4.2$

13.
$15.36 \div 3.2 = 4.8$

14.
$18.62 \div 3.8 = 4.9$

15.
$17.55 \div 3.9 = 4.5$

16.
$23.03 \div 4.9 = 4.7$

1.
$23.85 \div 4.5 = 5.3$

2.
$21.73 \div 4.1 = 5.3$

3.
$22.14 \div 4.1 = 5.4$

4.
$12.18 \div 2.9 = 4.2$

5.
$17.02 \div 3.7 = 4.6$

6.
$19.11 \div 4.9 = 3.9$

7.
$16.72 \div 3.8 = 4.4$

8.
$23.46 \div 4.6 = 5.1$

9.
$20.06 \div 3.4 = 5.9$

10.
$25.85 \div 5.5 = 4.7$

11.
$15.37 \div 2.9 = 5.3$

12.
$21.12 \div 4.8 = 4.4$

13.
$19.95 \div 5.7 = 3.5$

14.
$20.35 \div 3.7 = 5.5$

15.
$14.57 \div 3.1 = 4.7$

16.
$29.15 \div 5.3 = 5.5$

1.
$2.08 \div 0.8 = 2.6$

2.
$2.28 \div 0.6 = 3.8$

3.
$1.44 \div 0.3 = 4.8$

4.
$2.16 \div 0.4 = 5.4$

5.
$1.52 \div 0.4 = 3.8$

6.
$5.04 \div 0.9 = 5.6$

7.
$2.64 \div 0.6 = 4.4$

8.
$1.17 \div 0.3 = 3.9$

9.
$0.98 \div 0.2 = 4.9$

10.
$3.44 \div 0.8 = 4.3$

11.
$4.13 \div 0.7 = 5.9$

12.
$3.68 \div 0.8 = 4.6$

13.
$3.35 \div 0.5 = 6.7$

14.
$3.22 \div 0.7 = 4.6$

15. $2.4 \div 0.5 = 4.8$

16.
$2.24 \div 0.7 = 3.2$

1.
$4.76 \div 0.7 = 6.8$

2.
$3.95 \div 0.5 = 7.9$

3.
$7.11 \div 0.9 = 7.9$

4.
$2.88 \div 0.3 = 9.6$

5.
$3.12 \div 0.4 = 7.8$

6.
$2.52 \div 0.3 = 8.4$

7.
$2.94 \div 0.6 = 4.9$

8.
$4.55 \div 0.7 = 6.5$

9.
$4.83 \div 0.7 = 6.9$

10.
$4.41 \div 0.7 = 6.3$

11.
$5.84 \div 0.8 = 7.3$

12.
$7.04 \div 0.8 = 8.8$

13.
$3.04 \div 0.1 = 7.6$

14. $4.2 \div 0.5 = 8.4$

15.
$5.58 \div 0.6 = 9.3$

16.
$8.91 \div 0.9 = 9.9$

www.ingramcontent.com/pod-product-compliance
Lightning Source LLC
Chambersburg PA
CBHW080830220526
45467CB00008B/2243